Graph Convolutional Neural Networks for Computer Vision

Scrivener Publishing
100 Cummings Center, Suite 541J
Beverly, MA 01915-6106

Publishers at Scrivener
Martin Scrivener (martin@scrivenerpublishing.com)
Phillip Carmical (pcarmical@scrivenerpublishing.com)

Graph Convolutional Neural Networks for Computer Vision

Edited by

Malini Alagarsamy

Rajesh Kumar Dhanaraj

J. Felicia Lilian

Vandana Sharma

and

Gheorghita Ghinea

Scrivener Publishing

WILEY

This edition first published 2026 by John Wiley & Sons, Inc., 111 River Street, Hoboken, NJ 07030, USA and Scrivener Publishing LLC, 100 Cummings Center, Suite 541J, Beverly, MA 01915, USA

For more information about Scrivener publications please visit www.scrivenerpublishing.com.

Wiley Global Headquarters
111 River Street, Hoboken, NJ 07030, USA

For details of our global editorial offices, customer services, and more information about Wiley products visit us at www.wiley.com.

The manufacturer's authorized representative according to the EU General Product Safety Regulation is Wiley-VCH GmbH, Boschstr. 12, 69469 Weinheim, Germany, e-mail: Product_Safety@wiley.com.

Library of Congress Cataloging-in-Publication Data

ISBN 9781394356331

Cover image: Generated with AI using Adobe Firefly
Cover design by Russell Richardson

Set in size of 11pt and Minion Pro by Manila Typesetting Company, Makati, Philippines

Printed in the USA

10 9 8 7 6 5 4 3 2 1

Contents

Preface

Graph convolutional network (GCN) models are neural network architectures that can use the graph structure to aggregate node information from neighbors in a convolutional manner. Graph convolutional networks have obtained higher performance in a wide range of tasks and applications because of their high interpretive capability in learning graph representations. Many real-world problems involve more complex data structures where elements are related in non-grid ways. Graph Convolutional Neural Networks (GCNNs) are powerful tools designed to handle such graph-structured data. In contrast, graph convolutional networks have been used to address several computer vision issues with equivalent or even superior results. They capture relationships and dependencies between data points, making them useful for a wide range of visual tasks. This book introduces the strengths of GCNNs in solving various computer vision problems. It combines both theory and practical applications to give readers a complete picture. The first chapter explains the role of GCNNs in vision systems and how they enhance image understanding. One important application is scene graph generation, where GCNNs model the relationships between objects in a static image. Another chapter focuses on how CNN results can be transformed into graph data for node classification and edge prediction.

The book focuses on the comparison GCNNs with traditional CNN based techniques by analysing the recent research trends and open challenges. In the next chapter the graphs based models contribution for better visual understanding, replacing or improving the classical architecture of CNN are presented. Their use in combinatorial optimization problems, such as object matching and tracking, is also covered. GCNNs in the field of bioinformatics helps to predict molecular fingerprints in chemical and pharmaceutical research. Detecting of moving objects in videos is discussed and how spatio-temporal graphs are analyzed to forecast traffic flow, which is useful in smart city planning is explained in the next chapter. The book also focuses on GCNN in healthcare in supporting diagnosis tasks such as detecting chest diseases from X-rays, improving visual QA systems and also in analyzing molecular structures for drug discovery. The book covers zero-shot learning and knowledge graphs that helps to recognize images of unseen classes. Dynamic graphs, which change over time, are also explained through real-world case studies, showing how GCNNs adapt to new data.

Each chapter presents real-world examples and cases that are easy to follow. This makes the book suitable to illustrate each application, making the text appealing and accessible to both practitioners and students. It offers great insights into how GCNNs can solve complex vision problems by understanding structure and context thus makes it useful for students, researchers, academicians and professionals seeking more in-depth knowledge of graph-based models. With the rising importance of graph learning, the GCNN will likely serve as a fundamental tool in advanced intelligent vision. The reader of this book will have a comprehensive understanding of what is GCNN and its application on computer vision.

1

Role of Graph Convolutional Neural Networks (GCNN) in Computer Vision Applications

A. Malini[1]*, Vandana Sharma[2], J. Felicia Lilian[3], Rajesh Kumar Dhanaraj[4], Sharangapriyan S.[3] and Shrinivas S.[3]

[1]School of Computer Science and Engineering, Vellore Institute of Technology, Chennai, Tamil Nadu, India
[2]Christ University, Bengaluru, India
[3]Department of Computer Science and Business Systems, Thiagarajar College of Engineering, Tamil Nadu, India
[4]Symbiosis Institute of Computer Studies and Research (SICSR), Symbiosis International (Deemed University), Pune, India

Abstract
Graph Convolution Neural Networks (GCNNs) are an important concept in advancing computer vision by transforming the understanding and modeling of graph-structured data. They have a unique capability to capture intricate relations along with the visual content that goes beyond the traditional and usual convolutional neural networks, it also empowers computers to observe and interpret the complex interconnection between the elements in images, which enhances the depth and nuance of visual dentata analysis. As a revolutionary study in computer vision, GCNNs are poised to transform various industries by unleashing new frontiers in the visual information domain's analysis and interpretation. Their multifaceted applications promise to reshape the landscape of computer vision.

***Keywords*:** Graph convolutional neural networks (GCNNs), computer vision, graph-structured data, visual data analysis, visual information analysis

**Corresponding author*: a.malinicse@gmail.com; a.malini@vit.ac.in

Malini Alagarsamy, Rajesh Kumar Dhanaraj, J. Felicia Lilian, Vandana Sharma and Gheorghita Ghinea (eds.) Graph Convolutional Neural Networks for Computer Vision, (1–20) © 2026 Scrivener Publishing LLC

1.1 Introduction

In our rapidly changing universe, computer vision's sphere is undergoing an amazingly unusual growth transforming several areas and enhancing human life's standards significantly. This paradigm change is supported by the huge expansion of Profound Learning and Nervous Networks in the computer vision realm.

Unlike humans, who reflexively gather footage and images, computers are carefully designed and trained to understand data enabling sophisticated analysis and rapid decision making. This innovative path has extensive implications for several fields, including the progress of autonomous vehicles, advanced medical testing, and social media recommendation system enhancement.

In today's world, GCNN has many applications. In social media analysis, GCNNs can analyze using nodes and edges as social networks by modeling users and their interactions, improving suggestion systems, class detection, and studies on influence propagation and predicting atomic behavior, molecular properties, and interactions by molecule models as graphical structures are crucial for genomics research in bioinformatics. For autonomous vehicles, GCNNs can model the environment as a graph, where nodes represent objects, and edges represent spatial relationships allowing for better scene understanding and decision making.

Additionally, GCNNs are employed in 3D computer vision tasks, such as shape analysis and 3D reconstruction, where 3D objects are represented as graphs of connected vertices. In conclusion, Graph Convolutional Neural Networks are revolutionizing the way complex data structures are analyzed and processed expanding the horizons of what is achievable in computer vision and beyond. Their ability to leverage the power of graphs opens up new possibilities for innovation and application across various domains further enhancing the impact of computer vision technologies on our everyday lives and enhancing human life's standards significantly.

1.2 Understanding Convolutional Neural Network in Computer Vision

Space Invariant Artificial Neural Networks have been the spearhead of reform in computer vision, which is the foundation of many refinements in visual analysis and interpretations. It has a profound effect on object recognition and scene understanding. CNN is often referred to as ConvNets representing a confined type of artificial neural network originated to operate mesh-like data such as images and videos. They have a very long history heading back to the 1980s, with seminal efforts by researchers like Yann LeCun, Geoffrey Hinton, and Yoshua Bengio. However, CNN came to action in the 2010s largely due to advances in computational power, the availability of large labeled datasets, and improved training algorithms.

1.3 Core Components of CNN

1. **Convolutional Layers:** These layers apply filters (also termed as kernels) to the feed-in data to pull features. These filters are accustomed to extract the computing dot in every position, and it captures the local patterns available in the given input data.
2. **Layers for Pooling:** Pooling layers lessen the geometrical extents of the characteristic map produced by convolutional layers. General pooling functions consist of average pooling and max pooling, which retain the most important data while reducing the computational complexity.
3. **Connected Layers (Fully Connected):** The layers that are fully connected are considered standard neural networks that connect all axons from one layer to another typically leading to the final output. They serve as classifiers or regressors in the environs of computer vision assignments.

1.3.1 Hierarchy Feature Learning

One of CNNs' key advantages is its capacity for hierarchical learning from unfiltered pixel data. Higher layers combine these properties into more complicated representations as the data traverses through the network allowing the network to eventually recognize objects, patterns, and abstract concepts. Lower layers collect basic elements like edges and corners.

1.4 Extending CNNs to Handle Graph-Structured Data

For their remarkable performance in grid-like data, CNN has been utilized for a long time, especially in computer vision applications involving photos and videos. Graph-structured data, which has its own special set of difficulties and potential, is included in the broad landscape of data kinds, which goes far beyond grids.

1.4.1 Graphs—A Universal Data Structure

For data with complicated linkages, graphs provide a flexible representation. Each node can carry important information, and they are made up of vertices and the edges that connect them. Graphs are used in many different sectors, such as social media networks, biology, transportation, recommendation engines, and more. To derive useful insights from graphs, it is essential to interpret and analyze the patterns inside graph-structured data.

1.4.2 Challenges in Processing Graphs with CNN

1. **Irregular Data Structures:** Graphs do not have homogeneous grids like ordinary grids do, which makes it challenging to extract the crucial data from the input.
2. **Variable Neighborhoods:** Traditional CNNs struggle to handle the variable-sized neighborhoods that nodes in a graph can have.

1.4.3 Graph Convolutional Neural Networks (GCNNs): Bridging the Gap

To effectively extract features and patterns from structured data, GCNNs expand the concepts of CNNs to the world of graphs. By using graph convolutions that operate on nodes and their neighboring nodes, they provide an advantage by capturing the contextual information required for various applications.

1.4.4 Architectural Components of Graph Convolutional Layers

At the heart of GCNN is the representation of the input graph. This typically involves encoding both the graph structure and node attributes. Common representations include the following as given in Figure 1.1.

1. **Adjacency Matrix:** The adjacency matrix, often referred to as A, is a key data structure that encodes the connectivity of nodes in a graph. In a plot with N nodes, the adjacency matrix is an N × N matrix, where each element (i,j) indicates whether there is an interrelation between node i, j. Typically, it is a binary matrix where the value of 1 signifies the presence of an edge, while a value of 0 indicates no connection.

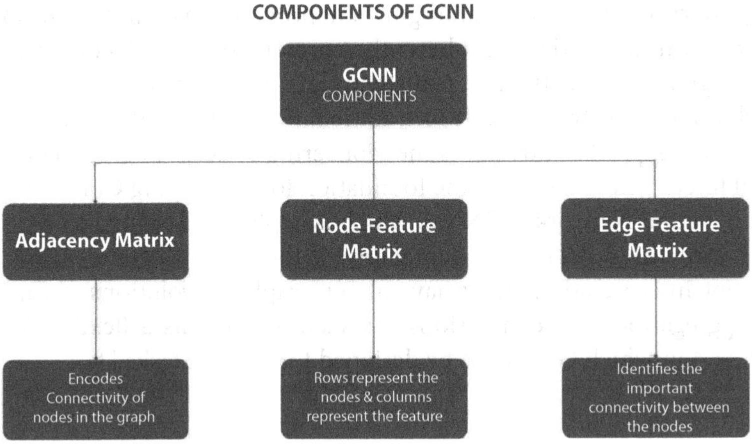

Figure 1.1 Components of GCNN.

2. **Node Feature Matrix:** This is a matrix in which each row indicates a node, and each column indicates an attribute. The node feature matrix is a structured data representation where each row conforms to a node in a graph, and each node represents the specific feature or attribute associated with that node. In essence, it acts as a table where nodes are entities, and the attributes are the properties of characteristics of those nodes.
3. **Edge Feature Matrix:** If the edge features are available, they can be encoded in a separate matrix. Edge feature matrices are responsible for enhanced connectivity between the nodes, which will give important information that is present in between the nodes. These added facts can reproduce various particulars of the relationships between nodes.

1.4.5 Graph Convolutional Layers: Adaptation of Convolutional Layers

The convolutional neural networks (CNNs) are famous for their effectiveness on structured data like time signals and images. They rely on the core operation of convolution with learned filters to extract meaningful features from input data. However, intrinsic limitations come with CNNs specifically as suitable only for regular structured data sources. This drawback restricts their use in cases where data have irregular or non-uniform patterns. In response to this limitation, Graph Signal Processing has emerged as a powerful replacement paradigm. GSP provides an entirely new way of dealing with signals that are inherently represented by graphs or networks. Such signals inherently encapsulate irregular and complex structures, which are not easily amenable to traditional convolutional operations. To bridge this gap, GSP introduces the pioneering concept of graph convolutional filters. These filters lay the foundation for developing Convolutional Graph Neural Networks (GNNs), which is explicitly designed to work on graph-structured data. The adaptability of these GNNs comes from two different interpretations they have about graph convolutions: diffusion- and aggregation-type conversions. As such, it presents a flexible framework within which GNNs can be designed for specific tasks [1].

1.5 Application of GCNN in Computer Vision

GCNN, which operates on graph-structured data, plays a pivotal part in the field of computer vision. Tasks that involve complex relations can be easily solved using it, especially for analyzing and understanding structures and pointing layers within the images. Some of its application are explained below.

1.5.1 Semantic Segmentation

Semantic segmentation is a process of dividing an image into classes and labeling them by training a model that has a main goal to locate and make a boundary around the objects. It is based on initializing every pixel in an image using a specific patch. Semantic segmentation involves multiple steps starting from image processing to visualization. It has lots of applications such as medical image analysis, understanding the scene, robotic perception, video security, augmented reality, and image condensing. Semantic segmentation uses the range of evolving efforts in instance and semantic segmentation, including pixel labeling using convolutional networks, encoder–decoder architectures, pyramid-based approaches and multiscale, recurrent networks, and non-human primate visual attention models [1].

1.5.2 Object Detection and Localization with GCNN

Object detection and localization using GCNNs involves representing an image as a graph, with nodes and edges denoting relationships. In computer vision, the relationship between object detection and graphs lies in how we represent and analyze images for the purpose of identifying objects within them. In deep learning, YOLO (You Only Look Once) processes an image by dividing it into a grille and predicts media boxes and class probabilities directly within each grille cell. YOLO's architecture allows for efficient real-time object detection by simultaneously processing the entire image in a single forward pass. It balances speed and accuracy making it suitable for various applications like autonomous driving and object tracking [2].

1.5.3 Graph-Based Image Classification

Image classification is an elementary assignment in computer vision, where the end result is to classify an image based on its visual content. For example, we can train an image classification algorithm to answer if a car is present in an image or not. While detecting an object is minor for humans, tough image classification is still a challenge in computer vision applications. It is a proposal to construct images as graphs and to try machine learning algorithms that operate on graphs to the emerging representations. Notably, the present approaches for representing images as graphs are dissimilar, and we endow on deep-rooted neural network architectures developed for graph-structured data to deal with image classification tasks. The proposed models leverage properties innate in images such as stationarity of statistics and locality of pixel dependencies. The proposed models in image classification tasks are compared with standard GCNN architectures.

1.5.4 Point Cloud Analysis and 3D Object Recognition

Deep learning methods are becoming more advanced for 3D object identification and categorization in self-driving applications for autonomous vehicles. A self-driving car system is reliant on the task. The availability of 3D scanners, such as LiDAR, in sensor technology produces more precise data. Newer versions of how LiDAR data works have recently undergone some significant development. High-definition LiDAR laser scanners capture the majority of 3D point clouds [3]. Additionally, by adopting a non-supervised ground and segmentation process to reduce scene complexity, it decreases the need for manually labeled training data [4].

1.5.5 Video Understanding and Action Recognition through Graphs

Videos address the issue of detecting human behavior. It is a graph-based structure to learn the way interactions occur between peoples and objects. It analyzes issues with spatiotemporal action perception and mundane action detection. Given a clip, the approach is to localize each actor and classify his actions. Both the actor and object has a space and time relationship between them. A representation of an actor and object exists as nodes and edges in a graph. Since there are significant correlations between actors and objects, the scene is represented as a fully connected graph, where every node is related to every other node and serves as the input for our neural network. The graph-based description is further subdivided into an attention layer and an adjacency matrix for segmentation to handle the consecutive clips acquired from the action detection of human behavior [5].

1.6 Enhancing Performance and Interpretability with GCNN

Interpretability: It is a long standing limitation of deep neural networks for classification results. It is the most straightforward approach for generating a sensitivity map over input data to discover the gradient map within the substructures. The gradient maps are noisy, and smoothing these maps by enhancing the performance of the interpretability of GCNN is necessary.

Enhancing the performance and interpretability of GCNNs is crucial for their broader adoption in applications that require both accuracy and transparency, such as social media network analysis, recommendation engines, and biology. Balancing performance and interpretability is an ongoing challenge, as more complex models might achieve better performance at the cost of interpretability. It is essential to choose the right techniques and strategies that align with your specific goals and constraints in any given application.

1.6.1 Handling Graph Irregularities and Noisy Data

Thresholding is fundamental for improving the performance of higher-level computer vision tasks including segmentation, classification, and object recognition. When the data are visualized as graph vertices, the convolution operation is converted into a generalized graph convolution. Denoising has always been a significant study area for developing deep learning systems when the graph is irregular. When the information state is structured as a graph and the data are represented as nodes in the graph, prolonging the GCNNs signals with a systematic form, such as images and videos, can produce a graph-structured signal. The graphs are actively computed in feature space utilizing GCDN architecture and graph convolutional layers [6].

1.6.2 Transfer Learning and Pre-Trained GCNN Models

Transfer learning often starts with pre-trained models, which are typically huge neural networks learned on massive datasets. These models have been trained to recognize and extract useful data features. The idea is to apply what you learned from the source task to improve your performance on the target task.

1.6.2.1 Transfer Learning Model for Waste Classification

This is a new combination classification model that processes the ImageNet database and achieves high classification accuracy using pretrained CNN models. The transfer learning model, based on each pretrained model, is built as a candidate classifier, and the best output from the CNN pretrained model's candidate classifiers is chosen as the final classification outcome [7].

1.6.2.2 Pre-Trained GCNN Models to Identify Finger Vein

CNN is the technology that is redefining the field of computer vision research, and it continues to advance accurate image classification in addition to being essential for performing tasks involving feature extraction and recognition, including image retrieval, object detection, and semantic segmentation. Because of the present era's technological advancements, which include improved computer power and GPU acceleration, CNN may be used robustly and effectively [8]. AlexNet, a model of two fully connected layers, five convolutional layers, and a softmax output layer, is based on deep CNN. This network is coherent in contrast to other, more rudimentary networks that use Rectified Linear Units, dropout, and overlap pooling.

1.6.3 Addressing Overfitting and Generalization Challenges

The amount of good solutions appears to be incompatible with the observed overfitting solutions when training deep neural networks. After training a deep neural network model on labeled data, it is typically tested on unlabeled data for evaluation. The model's capacity to generalize implies that it is capable of performing well on test data. Overfitting occurs when a model performs well on training data but fails on test data. More particular, the model learns the noise patterns contained in the training data resulting in a substantial difference between training and test error due to overfitting. Underfitting occurs when the model fails to catch patterns in both training and test data [9].

1.7 Future Directions and Emerging Trends

Graph Convolutional Neural Networks (GCNNs) have made enormous advances in the last decade and remain an important area of research. Among the future directions and developing developments for GCNNs are the following:

Improving Efficiency: Deep learning has always been a problem, especially when it comes to GCNNs. This is because scientists are now in search of ways to make GCNNs more efficient in terms of computation and memory. These techniques can be categorized as sparsity, quantization, and model compression.
AutoML for GCNNs: With the invention of GCNNs, automated machine learning tools (AutoML) have become more popular. AutoML is helping, automatically configuring, and optimizing GCNN models for specific applications.
Graph Neural Architecture Search (GNAS): GNAS involves finding the best architectures for specific tasks in AutoML. It saves time spent on architecture design and resources used thereof.
Graph Adversarial Attacks and Defenses: Since there are security applications of GCNNs in fraud detection, current investigations concentrate on graph aspects related to adversarial attacks and defenses. Among them is developing robust models against adversarial attacks targeted at defined systems using Graph Convolutional Networks (GCNs).

1.7.1 Advances in Graph Neural Network Architectures

1.7.1.1 Graph Attention Network

The Graph Attention Network introduced a mechanism for learning node-specific attention weights enabling GNNs to focus on different neighbors when aggregating information.

A GAT is a type of neural network design that works with graph-structured data. It generates methods based on graph convolutions using self-attention layers. Deep networks can be built by stacking GAT layers on top of each other [10]. The model can learn more abstract representations of nodes by allowing them to attend over their neighborhood's features multiple times by stacking layers. Through this stacking, the model is able to capture complicated linkages and hierarchies in the network.

1.7.1.2 GNNExplainer: Generating Explanations for Graph Neural Networks

GNN is a significant technique for machine learning on graphs. It integrates node feature information with graph structure by recursively propagating neural messages along the edges of the input graphs. GNN explainers are comprehensible explanations for GNN predictions. The use of this explainer increases trust in the GCNN model, transparency, and allows us to understand the network in it [11]. Given a trained GNN model, GNNEXPLAINER attempts to uncover the underlying elements driving these predictions. We have distinct kinds of predictions as follows: single-instance explanation and multi-instance explanation prediction.

1.7.1.2.1 Single-Instance Explanation

For single-instance explanations, GNNEXPLAINER identifies a subgraph within the computation graph and selects a subset of node features that are most influential in determining a specific prediction. This process allows us to gain insight into the critical components affecting individual predictions.

1.7.1.2.2 Multi-Instance Explanation

In scenarios where we need to explain a set of predictions, GNNEXPLAINER aggregates explanations from each prediction within the set. It then automatically summarizes these individual explanations into a prototype explanation providing a holistic view of the features and subgraphs that collectively influence the entire set of predictions [12].

1.7.2 Integration of Graph Structures with Traditional CNNs

The integration of graph structures with traditional CNNs has took to the development of Graph Convolutional Networks (GCNs) and other graph-based deep learning architectures. These models enable the processing of graph data while respecting the data's intrinsic structure making them applicable in various domains where graphs are a common representation.

1.7.2.1 Locally Connected and Spectral Networks on Graphs

- **Graph-Structured Data**
There are many real-world datasets that are typically represented as graphs, where the nodes represent entities and the edges that denote relationships or connections between these entities.

- **Spectral Graph Theory**
Spectral Graph Conjecture is one thing of arithmetic that queries the rudimentary peculiarities of diagrams utilizing linear arithmetic as an aperture. It exerts the links within the form of a diagram and its phantasmal attributes, especially the eigenvalues and eigen-drones of the diagram's proximity or Laplacian matrices [13]. The spectral measurements give valuable details about the connections between the nodes of the graph. It has clustering tendencies and community structures. Spectral Graph Theories have applications in analyzing networks, segmenting images, clustering data, and creating recommendation systems.

- **Deep Locally Connected Networks on Graphs**
DLCN on Graphs are a new-fashioned way to employ deep learning methods, especially Convolutional Neural Networks (CNNs), to graph-structured data. By contemplating graphs as squares, these networks authorize the modeling of non-grid facts giving a frame for applying convolutional processes to graph-structured datasets. Deep local connect networks, not like average CNNs, take a less stiff approach where every node may have an odd set of neighbors. This allows for learning of localized and non-Euclidean traits on graphs, which can be handy in various fields like social network scrutiny, molecular chemistry, and recommendation systems. These networks have unlocked fresh possibilities for tackling complex problems using non-grid, interconnected data by broadening deep learning activities to graph details.

1.7.3 Explainable AI in GCNNs for Computer Vision

Gradient maps have been used with GCNN to describe the plainability procedures in Explainable AI. Every graph is indicated as a node, where each node is represented as an attributed graph for classification purposes. In this graph, the node functions as a matrix, and the adjacency matrix encodes the node's connectedness. We acquire scalars over nodes when the explainable AI algorithm is applied to all samples. Using explainable AI techniques, the following metrics are computed from the provided samples: fidelity, contrastivity, sparsity, and heatmap [14].

1.8 Challenges and Open Research Questions

There are a number of urgent problems and unanswered research topics with graph convolutional neural networks (GCNNs). First, a basic problem remains: how to make GCNNs scalable on massive graphs with millions or billions of edges and nodes. The problem of adapting GCNNs to dynamic graphs, especially ones that change over time, and diverse data is complex. For practical applications, it is important to improve the explainability and generalization capacity of GCNNs, particularly for graphs with little labeled input.

Robust regularization strategies are necessary to prevent overfitting when there is small or noisy graph data. Enabling GCNNs for inductive learning, where they can generalize to unseen nodes or graphs, and exploring unsupervised learning approaches are open areas of investigation. Furthermore, addressing adversarial attacks and developing defenses is critical, especially in security-critical applications. Scalable architectural search, graph generative models, and the potential of quantum computing to accelerate graph-based computations are emerging research directions with significant challenges.

1.8.1 Scalability of GCNNs to Large Graphs

Graph Convolutional Neural Networks (GCNNs) face fundamental challenges in the field of graph-based deep learning: their scalability to huge networks. Countless numbers of nodes and edges can be found in large graphs, which presents severe memory and processing demands.

1.8.1.1 Scalable Graph Convolutional Networks

For handling the plot-structured data, GCNN with a swift sectional spectral filter for directed graphs is used. Two graphing methods currently in use are the spatial method, which focuses on each node's neighborhood, and the spectral method for graphs based on Laplacian. The majority of current methods operate on undirected graphs, which results in information loss [15]. To avoid this, one can use graph Laplacian, commonly known as a swift directed acyclic graph convolutional network, which operates on directed graphs. They use spatial linear aggregation and operate entirely on directed graphs.

For semi-supervised node classification, a fast directed graph convolutional network model, known as FDGCN, is used. The operations for a directed graph are precise, and node categorization is developed into a convolutional model based on the operations [16].

1.8.2 Robustness of GCNNs Against Adversarial Attacks

In applications requiring graph-structured data, the robustness of Graph Convolutional Neural Networks (GCNNs) against adversarial attacks is essential. Adversarial attacks entail changing the graph's structure or node attributes in such a way that the GCNN's predictions are misled potentially resulting in security breaches, misinformation, or incorrect recommendations. The creation of robust training algorithms and protection mechanisms, as well as recognizing the vulnerability of different GCNN architectures to various attack tactics, are open research problems in this domain. The defense techniques used, such as adversarial training, resilient pooling algorithms, or outlier detection, are determined by the application and threat model.

1.8.2.1 Understanding Adversarial Robustness of Symmetric Network

Furthermore, adversarial robustness—which is known to be vulnerable—is achieved by neural network models. Adversarial attacks are deployed on group-equivalent convolutional neural networks that are rotation equivariant.

1.8.3 Combining Graph-Based and Spatial Feature Representations

In deep learning, combining spatial and graph-based feature representations is a potent strategy, especially for jobs involving grid-based data, like pictures, and structured data, like graphs. By combining the best features of each representation, the model performs better overall.

1.8.3.1 *Analysis of Graph Convolutional Networks and Recent Datasets for Visual Question Answering*

GCNN is used in non-structural contexts because of its interpretability and significant performance. Graph reasoning models are utilized in visual question answering to develop a response system for an image provided as input, while simultaneously identifying the image's semantic meaning. A GCNN interpretability model is used for this question-answering system.

1.8.4 Ethical Considerations and Fairness in GCNN Applications

Ethical considerations and fairness in Graph Convolutional Neural Network (GCNN) applications are crucial aspects of deploying these models responsibly [17]. While GCNNs offer powerful tools for data analysis and decision making, they can also raise ethical concerns and introduce biases. Here are key ethical considerations and fairness issues in GCNN applications:

- Data Bias
- Transparency
- Explainability
- Fairness
- Bias Mitigation

1.8.4.1 *Neutrality and Abstraction in Sociotechnical Systems*

The GCNN program uses neutrality-aware learning algorithms to step in at various points in the decision-making process and ensure a just conclusion. This algorithm uses decision-making mechanisms and surrounds itself with traps to achieve fairness in sociotechnical phenomena [18].

Artificial Intelligence and ML algorithms are in charge of decision-making processes in the transportation, healthcare, and many other sectors. These days, decision algorithms are prone to unfairness-based applications [19]. This can be avoided by employing machine learning algorithms to detect and measure fairness in a way that is fully and precisely processed. Fairness bias boosting mechanisms are examined and applied to the decision-making scenarios.

1.9 Case Studies: Real-World Applications

In computer vision, GCNN is used in numerous real-world applications and case studies are tabulated and shown in Table 1.1. It has a significant impact on every situation that arises mainly in computer vision is shown in Figure 1.2. A few instances from the real world where it has had a significant influence is as follows [20]:

I. **Image Segmentation:** Images can be segmented using GCNN, which can recognize and label various objects in an image. Applications like robotics, self-driving cars, and medical image analysis can benefit from that.
II. **Object Detection:** Object detection in photos and videos can be accomplished with GCNNs—applications like traffic monitoring, security and surveillance, and retail analytics can benefit from this [21].
III. **Pose Estimation:** GCNNs are useful for estimating object poses in both videos and images. Applications like virtual reality, augmented reality, and human–computer vision can benefit from that.

Table 1.1 Applications of GCNN in computer vision.

Case study	Description	Potential benefits from GCNN
Medical image analysis	• In medical image analysis context, GCNNs are used to learn features from medical images that are relevant to disease diagnosis. • A GCNN-based system developed by researchers at Stanford University has been shown to be more accurate than human pathologists in detecting cancer cells in breast cancer images	• Improvised accuracy • Reduced subjectivity • Increased efficiency • Reduced costs

(*Continued*)

Table 1.1 Applications of GCNN in computer vision. (*Continued*)

Case study	Description	Potential benefits from GCNN
Self-driving cars	• Self-driving cars are utilizing an array of sensors, various cameras, radar, and LiDAR gathering data surrounding them • Within the self-driving car scenario, graph data may visualize the connections amid diverse objects on the highway	• Robustness • Reduced latency • Improvised detection and tracking system
Robotics	• One of the crucial challenges in robot grasp is that it is hard to model the complex interactions between a robot grip and an object • Graph data can signify the relationships between the various parts of the robot's grip and an object to be grasped • This detail can be useful to predict the result of a trail and to plan grasping tactics that are likely successful	• Reduced planning time • Improvised grasping accuracy and success rate • Increased robustness
Environmental monitoring	• Spotting and tracking environmental hazards such as wildfires, oil spills, and illegal logging • Researchers at the University of California, Berkeley, have developed a GCNN-based system that can detect wildfires with high definiteness • Scientists at the Massachusetts Institute of Technology have used GCNNs to develop a system that can monitor water caliber instances	• Enhanced surveillance • High accurate prediction

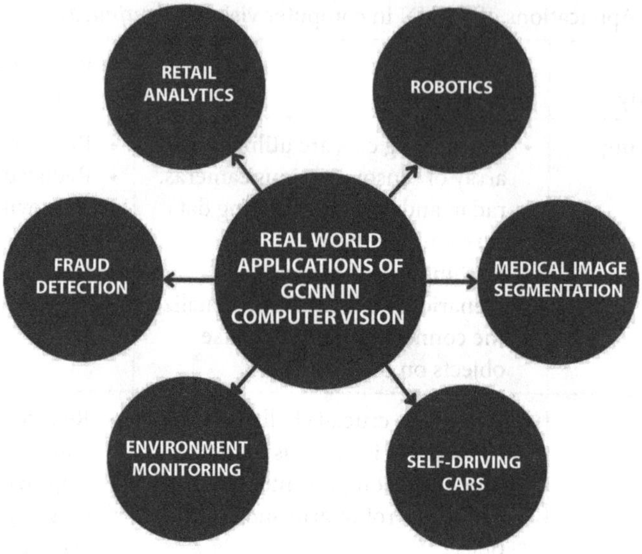

Figure 1.2 Real-world applications of GCNN in CV.

1.10 Conclusion

Graph Convolutional Neural Networks (GCNNs) are potent tools for computer vision tasks. GCNNs are capable of acquiring knowledge about complex features from graph-structured data, which makes them requisite for tasks such as object detection and image segmentation. GCNNs are still under development, but they have the chance to remodel the way that we perceive and interact with the world around us by improving the accuracy, efficiency, and robustness of computer vision, such as developing more intelligent and capable solutions and enhancing the reliability of anything that is related to graphical structure. There are a number of exciting areas of future research in GCNN-based computer vision such as developing GCNN-based systems that are more interpretable. This will help us to understand how GCNNs make predictions and identify potential biases in GCNN models. Finally, it is important to develop CNN-based systems that are robust to noise and uncertainty. This is important for real-world applications, where the data may be noisy or incomplete.

References

1. Minaee, S., Boykov, Y., Porikli, F., Plaza, A., Kehtarnavaz, N., Terzopoulos, D., Image segmentation using deep learning: A survey. *IEEE Trans. Pattern Anal. Mach. Intell.*, 44, 7, 3523–3542, 2021.
2. Dhillon, A. and Verma, G.K., Convolutional neural network: a review of models, methodologies and applications to object detection. *Prog. Artif. Intell.*, 9, 2, 85–112, 2020.
3. Basheer, S., Singh, K.U., Sharma, V., Bhatia, S., Pande, N., Kumar, A., A robust NIfTI image authentication framework to ensure reliable and safe diagnosis. *PeerJ Comput. Sci.*, 9, e1323, 2023.
4. Nikolentzos, G., Thomas, M., Rivera, A.R., Vazirgiannis, M., Image Classification Using Graph-Based Representations and Graph Neural Networks, in: *Complex Networks & Their Applications IX: Volume 2, Proceedings of the Ninth International Conference on Complex Networks and Their Applications Complex Networks 2020*, Springer International Publishing, pp. 142–153, 2021.
5. Babahajiani, P., Fan, L., Gabbouj, M., Object recognition in 3D point cloud of urban street scene, in: *Computer Vision-ACCV 2014 Workshops: Singapore*, Singapore, November 1–2, 2014, Springer International Publishing, pp. Revised Selected Papers, Part I 12 pp. 177–190, 2015.
6. Tomei, M., Baraldi, L., Calderara, S., Bronzin, S., Cucchiara, R., Video action detection by learning graph-based spatio-temporal interactions. *Comput. Vision Image Understanding*, 206, 103187, 2021.
7. Chen, Y., Huang, W., Zhang, D., Chen, W., An open-source Matlab code package for improved rank-reduction 3D seismic data denoising and reconstruction. *Comput. Geosci.*, 95, 59–66, 2016.
8. Zheng, J., Xu, M., Cai, M., Wang, Z., Yang, M., Modeling group behavior to study innovation diffusion based on cognition and network: an analysis for garbage classification system in Shanghai, China. *Int. J. Environ. Res. Public Health*, 16, 18, 3349, 2019.
9. Fairuz, S., Habaebi, M.H., Elsheikh, E.M.A., Pre-trained based CNN model to identify finger vein. *Bull. Electr. Eng. Inf.*, 8, 3, 855–862, 2019.
10. Salman, S. and Liu, X., *Overfitting mechanism and avoidance in deep neural networks*, 2019, *arXiv preprint arXiv:1901.06566*.
11. Veličković, P., *et al.*, *Graph attention networks*, 2017, *arXiv preprint arXiv:1710.10903*.
12. Ying, Z., *et al.*, Gnnexplainer: Generating explanations for graph neural networks. *Adv. Neural Inf. Process. Syst.*, 32, 2019.
13. Lucic, A., *et al.*, Cf-gnnexplainer: Counterfactual explanations for graph neural networks. *International Conference on Artificial Intelligence and Statistics. PMLR*, 2022.
14. Bruna, J., *et al.*, *Spectral networks and locally connected networks on graphs*, 2013, *arXiv preprint arXiv:1312.6203*.

15. Pope, P.E., *et al.*, Explainability methods for graph convolutional neural networks. *Proceedings of the IEEE/CVF Conference on Computer Vision and Pattern Recognition*, 2019.
16. Li, C., Qin, X., Xu, X., Yang, D., Wei, G., Scalable Graph Convolutional Networks With Fast Localized Spectral Filter for Directed Graphs. *IEEE Access*, 8, 105634–105644, 2020.
17. Kamath, S. and Deshpande, A., *Understanding Adversarial Robustness of Symmetric Networks*, 2018.
18. Yusuf, A.A., Chong, F., Xianling, M., An analysis of graph convolutional networks and recent datasets for visual question answering. *Artif. Intell. Rev.*, 55, 6277–6300, 2022.
19. Angwin, J., Larson, J., Mattu, S., & Kirchner, L. Machine bias, in: *Ethics of data and analytics*, pp. 254-264, 2022. Auerbach Publications.
20. Smith, B., Khojandi, A., Vasudevan, R., Bias in Reinforcement Learning: A Review in Healthcare Applications. *ACM Comput. Surv.*, 56, 2, 1–17, 2023.
21. Borkar, K.K., Aljrees, T., Pandey, S.K., Kumar, A., Singh, M.K., Sinha, A., Singh, K.U., Sharma, V., Stability Analysis and Navigational Techniques of Wheeled Mobile Robot: A Review. *Processes*, 11, 3302, 2023. https://doi.org/10.3390/pr11123302.

2

Scene Graph Generation from Static Images: Overview, Methods, and Applications

K. Krishnakishore[1]*, R. Vijayarangan[2], V. Jagan Naveen[3] and V. Kannan[4]

[1]*Department of Electronics and Communication Engineering, Vasavi College of Engineering, Hyderabad, India*
[2]*Department of Computer Science and Engineering, Sengunthar College of Engineering, Tiruchengode, Tamil Nadu, India*
[3]*Department of Electronics and Communication Engineering, GMR Institute of Technology, Rajam, Andhra Pradesh, India*
[4]*Department of Electronics and Communication Engineering, Bharat Institute of Engineering and Technology, Hyderabad, India*

Abstract
The process of creating scene graphs from still photos include a methodical analysis of visual situations *via* the recognition of elements and the connections between them. Objects are represented as nodes and their interactions as edges in a graph-like structure, with the goal of creating a complete representation of the visual content. Methods used in scene graph construction are covered in this review beginning with object recognition and ending with connection modeling. This study compares the effectiveness of many methods for inferring connections between objects in a picture. Visual question answering, picture captioning, recommendation systems, and robotics are just a few of the many areas that may benefit from scene graph creation. Caption generation, robotic navigation, and precise information retrieval are just a few examples of how image-related activities are improved by the deep scene knowledge made possible by these created graphs. Scalability, data annotation, and model interpretability continue to be areas of concern, notwithstanding the progress made.

***Keywords*:** Scene graph generation, object detection, relationship modeling, computer vision, graph convolutional networks (GCNs), visual understanding

**Corresponding author*: krishnakishore.k@staff.vce.ac.in

Malini Alagarsamy, Rajesh Kumar Dhanaraj, J. Felicia Lilian, Vandana Sharma and Gheorghita Ghinea (eds.) Graph Convolutional Neural Networks for Computer Vision, (21–46) © 2026 Scrivener Publishing LLC

2.1 Introduction

With the advent of deep learning and static picture recognition, higher-level visual understanding tasks, like visual connection detection, have exploded in popularity as research topics. Using a scene graph—an abstraction of items and their complex interactions [1]—rich semantic information may be gleaned from a static picture. It requires object localization and subject–predicate–object triplet identification in a static picture. A scene graph is a hierarchical picture representation that may facilitate many different kinds of high-level visual operations. Among these tasks are the following: producing static pictures, retrieving static images, answering visual queries, and captioning static images.

Our scene graph learning approach is conceptually shown in Figure 2.1.

In the green portion on the left, you can see how a static image is transformed into a scene graph. The blue area on the right shows the static image level regularize, which uses the item names and bounding boxes to recreate the static image. The top-level common sense knowledge reasoning is introduced throughout the scene graph creation procedure. Detecting and localizing pairs of interactive objects and recognizing their link to one another in a pairwise form makes the process of extracting scene graphs from static photographs not an easy one. Currently, scene graphs may be generated using one of two primary ways. Predicate inference is performed in both instances by grouping the object suggestions into pairs and then applying the phrase characteristics (features of their union area). What differentiates one category from another is the unique procedure that each category follows. It is only when the first category has identified the items that they begin to understand the relationships between them. To jointly ascertain the objects' identities and their relationships, the second category makes use of both object proposals and relationship proposals.

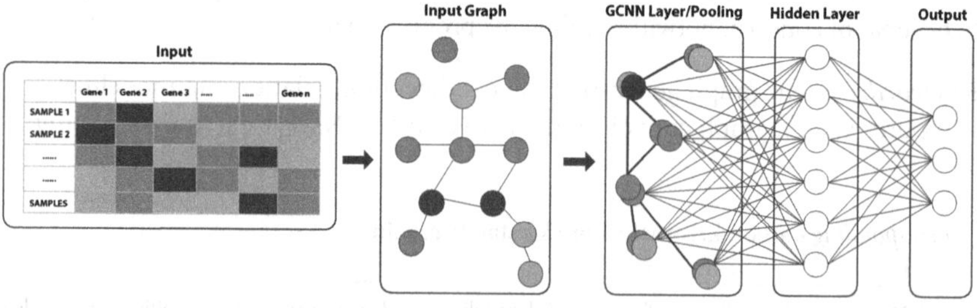

Figure 2.1 Conceptual illustration of our scene graph learning model.

Despite the promising advances made possible by these approaches, most of them are limited by the limitations of the current scene graph datasets. Before the scene graph can accurately depict a static picture, a varied set of connection triplets—subject, predicate, and objective—must be assembled [4]. Unfortunately, current datasets, like the Visual Relationship Detection (VRD) dataset, only capture a small fraction of the relevant information. Using a dataset that [2] has distributions with broad tails to train a prediction model will cause the model to favor those relationships that occur most often. Second, the construction of predicate labels is significantly affected by the identification of object pairs. Noises persist in current large-scale crowd-sourced datasets like Visual Genome (VG) because it is not feasible to precisely pinpoint the bounding boxes of every instance of each item (e.g., missing annotations and meaningless proposals). Trainee object detectors with poor performance are inevitable in a dataset with this much noise, which further hampers predicate detection performance.

The visual elements of a static picture may be deduced from our daily experiences by using our reasoning abilities as humans. Take Figure 2.1 as an example. It is clear that the subject, the woman, appears or stands on something, and the object, the snow, supports the predicate with proof (skiing). More precise object identification could potentially benefit from common sense knowledge. One example is how the specific external data needed for skiing aids in drawing conclusions about the item (snow). This motivates us to use intuitive data to ease the process of developing scene graphs. There is a lot of noise in existing datasets due to missing object annotations, even though object labels are crucial for link prediction. Still, we would want to get our hands on scene graphs that provide a more complete picture of the scenario. With this goal in mind, we have been working to regularize our scene graph generation network by using the identified components to recreate the static picture. Considering the situation shown in Figure 2.1, a method may mistakenly classify snow as grass. We would severely penalize this little error if we created a static image using the inaccurately predicted scene graph, even if most of the snow's relationships are correctly identified. There are essentially three types of contributions that this article makes as follows. 1) Based on our research, we propose a feature refining module that uses external sources of common sense information. In particular, before scene graphs are built, the module retrieves crucial data from ConceptNet to enhance phrase and object properties. To infer the most probable relations from the recovered data, we use Dynamic Memory Network, or DMN, for multihop reasoning. 2) After reconstructing the static image, we create a supervision module at the static image level to standardize the model for generating scene graphs. Since this auxiliary branch is only active during training, its primary function is to serve as a regularizer. 3) The VRD dataset and the VG dataset are two separate datasets that we use for extensive testing. Based on our experimental results, our approach may significantly improve upon existing scene graph generation techniques.

2.2 Definition

Figure 2.2 shows the overall process of building a scene graph, illustrating the steps from capturing the image to enhancing and processing it with a CNN model.

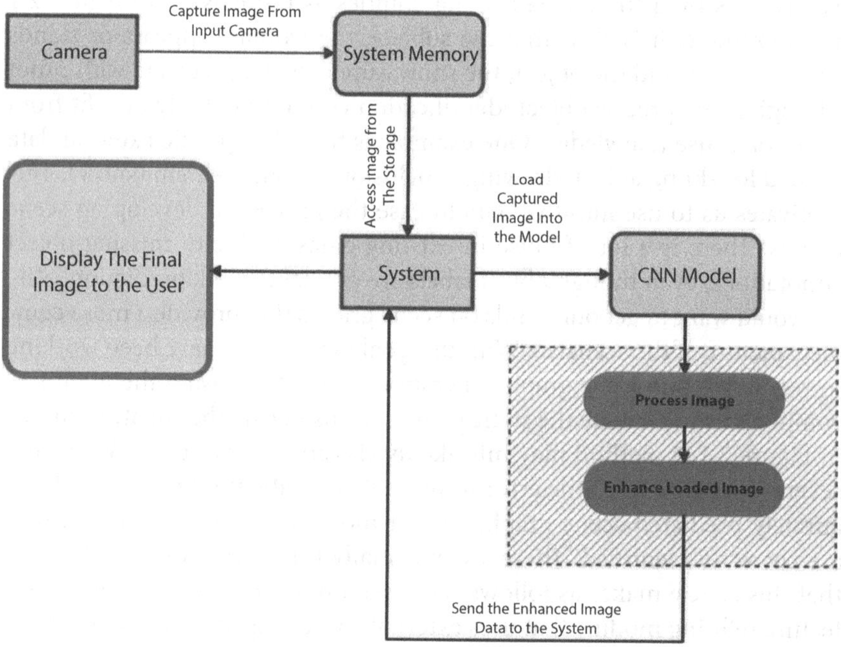

Figure 2.2 The overall process of building a scene graph.

2.3 Challenge

Nevertheless, there are still many challenges that academics dealing with scene graphs face. There are now two main problems with scene graphs that researchers are trying to solve as follows:

- Accuracy of the scene graph generation (SGG). Finding out how to make a completer and more accurate scene graph is the main problem now. The scene graph created by visual text information mining is heavily impacted by the quality and comprehensiveness of the learning models used. This highlights the critical importance of related learning model research for SGG.
- Knowledge sharing from past data. In addition to thoroughly mining the objects and their interactions in the current training set, more historical data are needed for the building of the scene graph. Making the most of the knowledge and skills that are currently accessible is another important dilemma.

2.4 Scene Graph Generation

Consequently, learning about related visual activities is greatly facilitated by the process of automatically producing a thorough and accurate scene graph. Scene graph creation approaches may be broadly classified into three types as follows: CRF based, TransE based, and CNN based.

2.5 Static Image

Poor lighting, noise, occlusion, and other factors may all lower the quality of static images while they are being captured. Geometric distortion and radiometric degradation are two of the most common kinds of difficulties in computer vision. Developing trustworthy pattern recognition methods that are insensitive to the many geometric and radiometric modifications that have been investigated has been the subject of much research effort. In this chapter, we have seen two methods for classifying multiclass objects in static photos that have been warped by geometrical transform, blur contamination, or both. The original method relied on wavelet features to categorize data. As with the second approach, it involves applying spatially symmetric blur-invariant descriptors. These are affined invariant moment descriptors and mixed blur descriptors.

To demonstrate the system's efficacy, many experiments are used. To determine which feature extraction methods work best, three independent tests are conducted. Object classification in static pictures after geometric modification, often referred to as an affine transform, is the primary subject of the first experiment. Regarding the second experiment, blur's impact on classification accuracy is examined. Static images with a combination of image degradation due to blurring and shape distortion due to affine transform are used to assess the system's performance in the end. Compared to a wavelet-based method, a moment-based strategy outperforms it in terms of effectiveness, processing power consumption, and applicability to static pictures with a wider range of geometrical and blur degradations. The sections that follow provide a breakdown of the remaining portions of the chapter. To explain how deteriorated static images behave, the following sections detail the static image degradation model and the static image deformation model. Next, we will go over two distinct methods for feature extraction, and then we will provide an experimental setup for classification models to illustrate how it works. Last but not the least, experimental data on how various degradations affect object classification are presented.

2.6 Degradation of a Static Image

The quality of the static pictures is the only factor that determines how challenging it is to identify and categorize them. Several factors contribute to the static image's diminished quality. The general degradation model's block diagram is shown in Figure 2.3.

In the parts that follow, we will talk about two different models of static image deterioration that were employed in this chapter. The following is a list of these:

2.6.1 Blurring Model

There are several potential causes of static picture blurring, which is among the most prevalent forms of image deterioration. These include, but are not limited to, diffraction, Jens aberration, incorrect focus, and air turbulence [3]. A loss of bandwidth, or blurring, occurs during the creation of a perfect static picture. This is because imperfection characterizes the creation process of static pictures. One probable explanation is that the optical system was out of focus, while another is that the camera and the

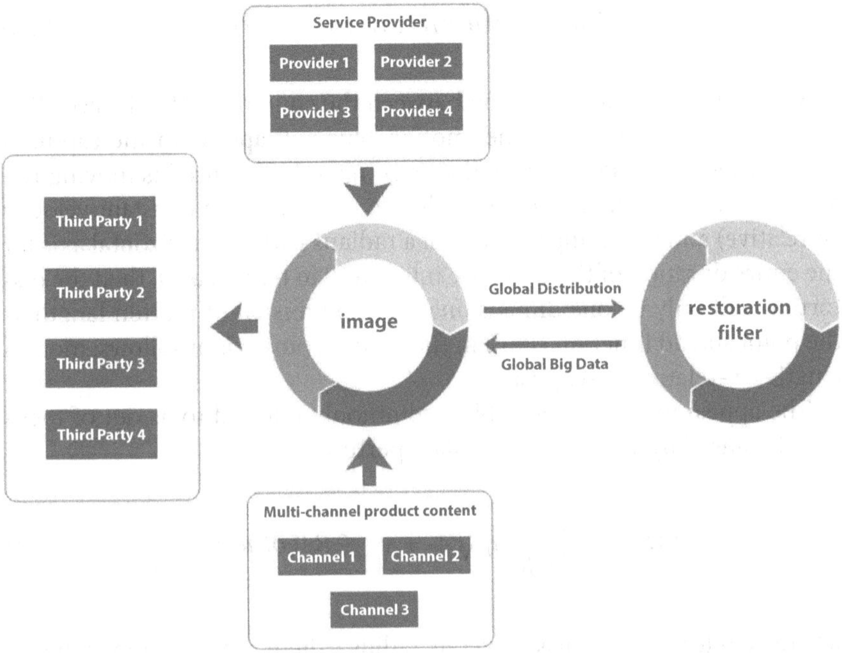

Figure 2.3 Static image degradation model.

scene being shot were moving relative to each other [5]. There are a lot of potential sources of blur in aerial photographs captured for remote sensing such as optical system aberrations, atmospheric turbulence, and relative camera-to-ground motion. Blurring of this sort is not limited to optical static images; for example, electron micrographs may be contaminated by spherical aberrations of the electron lenses. The deterioration process is another name for the blurring process. The convolution method described by Gonzalez and Woods may be used to depict this process.

$$i(x,y) = s(x,y) \otimes h(x,y) + n(x,y) \tag{2.1}$$

In this context, the sign $\otimes$ denotes a convolution that occurs in two dimensions. Convolution is used to illustrate the process of applying a blurring function to a static image. This means that one pixel of the blurred static image is created from one region of the original static image. The counterpart of the convolution operations specified in the spatial domain is the multiplication in the frequency domain, and the static image deterioration model is the result of this.

$$l(u,v) = S(u,v)H(u,v) + N(u,v) \tag{2.2}$$

When there is relative motion between the equipment that is recording the static image and the scene, motion blur will appear in the captured static image. If the static image that is going to be acquired is moving relative to the device that is going to do the capturing at a constant velocity (v-relative) and is making an angle of a radians with the horizontal axis for the entire duration of the exposure interval (0 to t-exposure), then the distortion will only be one dimensional. The expression of motion length as the point spread function (PSF) in the spatial domain is as follows: motion length = v-relative × t-exposure.

The appearance of defocus blur, commonly referred to as out of focus blur, is caused by a system of circular aperture.

$$h(x,y) = \begin{cases} \frac{1}{\pi R^2} if \sqrt{x^2+y^2} \leq R \, 0 \, otherwise \end{cases} \tag{2.3}$$

where R refers to the disk's overall radius. There are occasions when a co-existence of both blurs is present in the static image. In this scenario, the blur model becomes as follows:

$$h(x,y) = a(x,y) \otimes b(x,y) \tag{2.4}$$

where $a(x,y), b(x,y)$ are point spread functions for motions and defocus blur, respectively, and $\otimes$ is the convolution operator. The degradation process model Equations 2.1 and 2.2 can be expressed as Equations 2.6 and 2.7, respectively, as follows:

$$\begin{aligned} i(x,y) &= s(x,y) \otimes a(x,y) \otimes b(x,y) + n(x,y) I(u,v) \\ &= S(u,v)A(u,v)B(u,v) + N(u,v) \end{aligned} \tag{2.5}$$

2.6.2 Affine Transformation

Pattern recognition and computer vision are two fields that make extensive use of affine moment invariants. In most cases, the affine transform may be divided into a number of other geometric transformations. Calculating these transformations requires an understanding of their connections to the geometric moments. Affine transformations are those that maintain the collinearity after being applied, which means that all points that were previously lying on a line continue to lie on a line after the transformation, and distance ratios stay the same (e.g., the midpoint of a line segment remains the midpoint after transformation). Combining rotations, translations, scaling, and shears is what makes up an affine transformation, which may be expressed by the equations below.

$$u = a_2x + a_1y + a_0v = b_2x + b_1y + b_0 \tag{2.6}$$

where translation is impacted by a-0, and b-0, and scaling is affected by a-2 and b-1. A combination of these two creates rotations and shears.

A chain of simple affine transformations may be used as building blocks to create more complex affine transformations. Matrix operations are used as a definitional framework for transform combinations. The following is an expression in terms of a matrix that describes the affine transformations.

$$[u\ v\ 1] = [a\ b\ c\ d\ e\ f\ 0\ 0\ 1][x\ y\ 1] \tag{2.7}$$

An equivalent expression using matrix notation is

$$q = Tp \tag{2.8}$$

where q, T, and p are defined above.

2.7 Method 1: Wavelet Feature Extraction

Wavelets are mathematical functions that assist in the description of the initial static image into a static image in the frequency domain. This static image may then be further separated into sub-band static images consisting of various frequency components. When studying each component, a resolution that is proportional to its scale is used. When it comes to the analysis of physical conditions in which the signal involves discontinuities and strong spikes, the Wavelet transform provides benefits over the more standard Fourier technique. Wavelet transforms have been effectively used in a variety of applications over the last two decades. Some examples of these applications include feature extraction and categorization. A wavelet-based method that considerably improved detection and robustness in multiple object situations was given. Wavelet transformations and its variations, such as wavelet packets, have become an ideal starting point for categorization owing to their scale-space localization capabilities, particularly in the case of textural static images. When attempting to extract the characteristics of a texture from a static image, the filters that are used in the wavelet transform play a crucial role [6].

The wavelet transform relies on the application of fundamental functions such as scaling and wavelets. These functions are generated by the scaling and translating of a foundational function referred to as the mother wavelet and are described as follows:

$$\varnothing^{a,b}(x) = \frac{1}{\sqrt{a}}\varnothing\left(\frac{x-b}{a}\right) \tag{2.9}$$

where "a" is the scaling term and "b" is the translation factor. In 2-D, it can be represented as

$$\varnothing^{a,b}(x,y) = \frac{1}{\sqrt{a_1 a_2}}\varnothing\left(\frac{x-b_1}{a_1},\frac{x-b_2}{a_2}\right) \tag{2.10}$$

The wavelet transform of a static image $f(x,y)$ is a set of coefficients $c_{i,j}$ that depend on the scale and shifting of the signal defined as

$$c_{i,j} = \int_{-\infty}^{\infty} \Sigma f(x,y)\varnothing^{a,b}(x,y)dxdy \tag{2.11}$$

Methodologies for the extraction of features examine both the objects and the static images to identify the characteristics that are most prevalent and that best reflect the different types of static images. To extract the characteristics of the object utilizing static images and discrete wavelet transform, the following approach is employed.

Step 1: Read the RGB Static Image $f(x,y,z)$.
Step 2: Convert the static image to gray scale static image $f(x,y)$.
Step 3: Compute the Haar wavelet coefficients of Step 2 static image, which returns four matrices A, H, V, D corresponding to approximations, horizontal, vertical, and diagonal coefficients of static image.
Step 4: Six feature descriptors from three coefficient matrices horizontal, vertical, and diagonal are extracted as follows:

A. Feature (1)
The mean or the average is described as

$$x^{-} = \frac{1}{N}\sum_{i=1}^{N} \Sigma(x_i - x^{-})^2 \tag{2.12}$$

The standard deviation is as follows:

$$var = \frac{1}{N-1}\sum_{i=1}^{n} \Sigma(x,-x^{-})^2 \tag{2.13}$$

The standard deviation is defined as the square root of variance

$$\sigma = \sqrt{var} \tag{2.14}$$

Smoothness is measured with its second moment as follows:

$$Smoothness = 1 - \frac{1}{(1+var)} \tag{2.15}$$

As will be seen in the following, the skewness, also known as the third moment, is a measurement of the asymmetry of the distribution.

The amount of information that can be found on the internet has exploded in recent years thanks to developments in technology. In the past, the two publishing firms were responsible for generating the material. However, now that there is widespread availability of the internet, individuals themselves are able to contribute content *via* the usage of weblogs and microblogs like Facebook and Twitter. As a result, there is a significant need for mining the web and obtaining information that is relevant. Semantically driven mining approaches are required to enable automated relevance detection of the information to be mined. One of these techniques, known as concept-based mining, is one way in which semantic relations may be exploited in mining. In this approach, conceptual graph (CG) formalism is applied to define the concept and its relations. During this stage, the fundamental ideas must be uncovered, and then those ideas must be put to use to mine the pertinent data. The idea that was developed in this way might be used in a variety of contexts, including the retrieval of information, the extraction of information, the building of databases, and the generation of summaries. A conceptual graph, often known as a CG, is a bipartite network that is made up of two different kinds of nodes: concept nodes and relation nodes. It is a graph representation for logic that is based on the semantic networks of artificial intelligence as well as Charles Sanders Peirce's existential graphs.

In the course of his work on database interfaces, John Sowa was the first person to present CGs. It provided graphical representations of CGs that could be used to link natural language inquiries onto conceptual schemata. Each schema had a declarative component graph (CG) with actor nodes that connected to it and represented either database relations or functions. Concept nodes are said to represent entities, characteristics, events, and actions in accordance with Sowa's definition of CGs. Relation nodes are used to express the nature of the connection that exists between two idea nodes. The primary benefit of expressing natural language text in the form of CG is that CGs may be quickly transformed to any Knowledge Interchange Format, such as first-order logic, which aids in semantic processing. This is a significant time saver. The process of automatically extracting CGs from texts has a number of obstacles, one of the most significant of which is the identification of ideas. The example that is provided below will provide more understanding into CGs.

There are four distinct ideas at play in the example sentence that was shown earlier. These are the subject of the phrase, Marie, the action of hitting, the notion of a piggy bank, and the concept of a hammer.

The meaning of the statement is conveyed by it effectively. A blow from an instrument was delivered to the object piggy bank by the agent Marie. Concept is widely investigated within a variety of disciplines, and the definition of concept is thoroughly addressed within the fields of linguistics, psychology, and artificial intelligence, among others. In the next part of this article, we will provide an outline of how the concept of concept is handled in the aforementioned three categories.

2.8 Psychological Perspective

The psychological viewpoint is the culmination of previous research in cognitive and behavioral psychology, which was combined to form the synthesis. How are concepts or ideas represented in the brain? What are the processes of perception, memory, and learning? How does it impact or govern behavior? These are all questions that are of interest to psychologists working in the subject of psychology, and "associationism" is one of the oldest theories in the field of psychology. According to this theory, a feeling is linked to a concept, and that concept leads to another concept, which in turn leads to even more concepts.

Otto Selz established his idea of "schematic anticipation" [7]. According to this theory, the solution to a problem is not discovered by undirected association; rather, the answer is obtained by discovering the concepts to fill in the gaps of a partly finished schema. Experiments were carried out by psychologists to demonstrate the significance of conceptual thinking as well as reasoning based on static images. Cognitive psychology has largely overtaken associationism and behaviorism as the dominant school of thought in the field of psychology. Psychologists who specialize in cognitive psychology study topics such as the mind, intellect, thinking, and knowledge. A number of trials using Norwegian white rats to see how well they could navigate a labyrinth.

It was previously thought by behaviorists that the rat learns depending on the stimuli that it encounters along the many pathways of the labyrinth. But when Tolman repeated the same studies using a labyrinth that had been flooded, the rat was still able to accurately navigate its way around the maze, despite the fact that it had never been trained to correlate swimming movements with the stimuli. Tolman postulated that rats had cognitive maps, and that throughout the process of learning, these maps are formed in the rats' brains and minds. A concept is a mental grouping of comparable objects, events, or people that is used to remember and comprehend what things are, what they signify, and what categories or groups they belong to from a psychological point of view. Concepts are used to remember and grasp what things are. For instance, if someone says, "think of a rosagulla, a sweet," the word "sweet" will cause your mind to conjure up some concepts about what type of sweet rosagulla is and what components it has. Is this some kind of milk-based candy? You are attempting to classify rosagulla into the kind it is by analyzing its components, including its flavor, texture, and appearance [8].

2.9 Linguistic Perspective

In 2001, the authors in [9] advanced the hypothesis that ideas are conceived *via* the use of language. The words or phrases of a language are used for labeling or identifying the ideas, and by doing so, they differentiate the concept from other concepts that have been labeled in the past or that may be labeled in the future. When an information is processed at the intermediate level, we first become aware of the stimuli and then begin categorizing it. This happens whenever an information has a level value of 5. Therefore, not only does language make it possible for us to generate new ideas but it also assists us in giving these ideas names, which makes the process of storing memories easier and more streamlined.

An observation that having a smaller vocabulary is often associated with a lower level of comprehension or knowledge. For instance, the Inuit people who live in Alaska have a variety of phrases for snow, depending on how it looks, how it feels, and what it is made of. If Inuit people are able to tell the difference between various types of snow based on these characteristics, then it stands to reason that they have a deeper understanding of snow than the majority of Americans do. Additionally, there are a disproportionately small number of words in the Hopi language that refer to the past or the future; the majority of action words are built using the present tense. This illustrates how much more important the Hopi Indians consider the present to be than either the past or the future.

In conclusion, the Philippines is an agricultural country that relies heavily on rice as its primary source of nutrition. Contrast that with the wheat that is grown in western nations. As a consequence of this, they have an extensive lexicon pertaining to rice, categorizing it in accordance with terms such as cooked, uncooked, overcooked, undercooked, fried, leftovers, and a great deal more according to class or kind. This demonstrates that Filipinos are well versed in all matters pertaining to rice. As a result, both Whorf and his pupil reached the conclusion that having a larger vocabulary indicates a higher level of understanding. Accordingly, Filipinos have a greater understanding of rice, Inuits in Alaska have a greater knowledge of snow, and Hopi Indians, on the other hand, do not have a great deal of information of the past. Is it reasonable to draw the conclusion that a person has a lower ceiling for the kinds of ideas and information they can comprehend if they have a limited vocabulary?

Rosch argued that this was not the case. Eleanor Rosch challenged the claims made by Whorf and Sapir in 1973 by demonstrating that a lack of vocabulary on a topic does not always indicate a lack of expertise on that topic. White and black are the only terms that are used to describe people's skin tones in the Dani culture of New Guinea, for instance [10]. In spite of the Danis' inability to name colors, further investigation reveals that they are nevertheless able to differentiate between shades of the same color. Taking the preceding instances into consideration, this may indicate that the Inuits place the same level of importance on snow as the Hopi Indians do at present and that the Filipinos place the same importance on rice.

2.10 Concepts and Conceptual Structures in Artificial Intelligence Perspective

If you want to be a part of artificial intelligence (AI), you need to know semantics. That is because AI needs a theory, formalism, or representation that can help computers and people communicate. Consequently, semantic graph representations became quite popular. The original shapes were created [14] as a visual representation of symbolic logic; he called them existential graphs. In 1960, Peirce's writings were published. Dependency grammar graphs were created [11]. "Correlational nets" are a kind of graph that Silvio Ceccato found in 1961. To facilitate machine translation, this specific graph type was used. The realm of meaning-expressing graphs encompasses a wide variety of types, including "conceptual dependency graphs" and "semantic nets." Database schema construction and natural language querying are two applications of this method. The use of conceptual graphs to express meaning in AI systems is on the rise.

The sign or phrase is in the top left corner, the idea, point, object, or extension is at the top right, and at the peak is the concept, intention, thought, idea, or sense. An individual may have an idea of something but no word to describe it, or they may have a phrase for a concept but no idea what it stands for; in these cases, one of the three sides of the triangle can be missing (abstract concept). "Concept mining" describes the process of gleaning the concepts from textual content. An idea or object may be represented by a notion. The selection is decided solely depending on the interpretation of the sentence's semantics; it can be a word or a phrase.

As a result, AI capabilities, like machine learning and natural language processing, are necessary for concept recognition from text sources. Idea mining is a challenging procedure, and it is of the highest importance to identify the components of a concept. Identifying concepts in a text not only helps in understanding connections but also provides a semantic representation of the text, which is crucial for a thorough knowledge of the material. Our review of the literature indicates that using a dictionary, thesaurus, WordNet, or lexicon has long been considered the gold standard for elucidating conceptual concepts. The most common approach for concept identification is building a generative probabilistic model of a bag of words. Several non-parametric variants of Latent Dirichlet Allocation (LDA) and Latent Semantic Analysis (LSA) are examples of methods that use comparable methodology. Another common approach is to use feature learning to construct discriminative probabilistic models. Models like Conditional Random Fields (CRFs) and Support Vector Machines are common examples of this kind (SVMs). Research has shown that SVMs and CRFs outperform LSA and LDA. All of these machine learning methods need large amounts of data with tags or annotations, which is common information. Since there is a strict limit on the amount of labeled data that may be utilized, methods like unsupervised or semi-supervised data analysis must be used. An example of a machine learning methodology that does not need human supervision is deep learning. Although these methods make use of massive quantities of unlabeled data, they operate with much less labeled data. For "semantic" comparison, it has been shown that non-linear dimensionality reduction approaches may extract a better abstract level text representation. These results show that deep learning is a good fit for detecting abstract text projections. In this work, Restricted Boltzmann Machines are used (RBMs). One important feature of RBMs is that they are able to avoid directed graphical models because of their complementary prior over hidden units. Despite RBMs' accuracy in inferring the hidden units' states and their speed of training (e.g., by contrastive divergence), the model sets a strict, implicit prior.

Its primary functions include the definition, recapitulation, amendment, storage, and dissemination of information. Contrasting CGs with UNL, one implementation of a generalized human language is Universal Networking Language (UNL) [12]. Syntax and rules for the creation of grammatical structures are present in UNL, just as they are in any other language. In addition, UNL is mostly used as a language across languages.

The converse is also true: Conceptual Graphs are not formal languages. Since it leans more toward abstraction, it is not hard to transform it into first-order logic. It all started in 1996 with a project called the UNL Programme, which was initiated by the Institute of Advanced Studies of the United Nations University in Tokyo in Japan. A set of core software and tools necessary for the UNL System's functioning have been assembled into an English-language architecture. Afterward, these criteria were put into practice. Our investigation has led us to believe that Balaji has built summaries and semantic representations of text using UNL for Tamil. Their approach makes use of a UNL graph [13] that is built *via* a semi-automatic technique. Among the many potential uses for UNL, machine translation is one to be considered. The UNL has a predefined set of relationships. Building reasoning and inference engines on top of UNL is not an easy task.

2.11 Applications of CGS

Applications, like information retrieval systems and question answering systems, have included conceptual graphs to improve their functionality. A framework for quality assurance (QA) systems, "AnswerFinder," was included in TREC 2004. The graph teaches us about the patterns in the questions and answers. Their study relies on conceptual graphs constructed from translated versions of logical phrase forms included in the TREC 2004 training data set of questions and answers. The most extensive graph overlap serves as the basis for the matching procedure. Using this procedure, they measured an average mean reciprocal ratio of 24.97% and an average accuracy of 21.44%.

They depict objects by combining the VSM model with computer visuals. Specifically, the information retrieval process is divided into two parts. The first step is to gather the necessary papers using the VSM model. The generated papers are inputted into the CG model, which ultimately extracts the documents containing the most relevant information. Here, just a small set of 13 semantic relations are used to produce CGs. It is from these syntactic patterns that these semantic linkages are built. For these tests, the CACM-3024 data collecting system was used. They

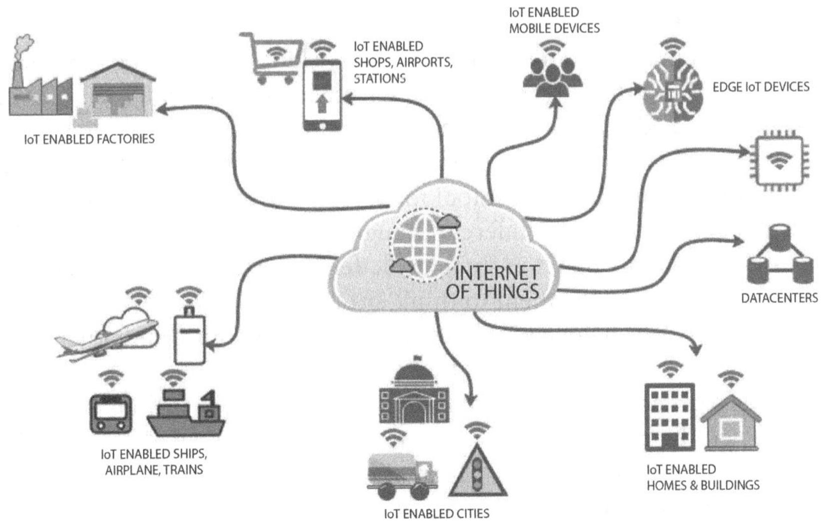

Figure 2.4 Overview of proposed work.

show a general increase in retrieval performance of 7.37% and an accuracy improvement of 34.8%. A study by Montes-y-Gomez *et al.* discusses information retrieval using CGs. Figure 2.4 presents the IoT-enabled application domains that benefit from conceptual graph models.

They provide a method for comparing two conceptual graphs as part of their study. Both the concepts and the relationships in the graph are considered for determining the degree of similarity using the dice coefficient, which is the basis for the similarity measure. The process of defining geographic information also makes use of conceptual graphs (CGs), in addition to their employment in knowledge base construction. They are responsible for creating a computer-generated model of each geographical need as part of their work. When all of the CGs that were constructed in the same way are linked together, they create a geographic knowledge base. All of the aforementioned research employed CGs for a wide range of tasks after retrieving the conceptual graphs, which may be done manually or semi-automatically. Here, heuristic principles are used to condense the long and complex sentences, and pattern matching criteria are used to check for the presence of concepts and relations (Shih). The rules are generated by a thorough and exhaustive text parsing. A recall rate of 70.2% and an accuracy rate of 78.75% were both accomplished by them. One of the biggest drawbacks is that it cannot be scaled, which might make it lose its purpose. In the realm of education, CGs have potential uses, particularly in the creation of resources to facilitate students' understanding of different concepts (as in concept-based learning). Participating in concept-based education equips children with the necessary skills to independently study the foundations. This strategy is both effective and efficient. The focus of concept-based education is on expanding students' understanding of various concepts rather than on teaching them to memorize facts by heart. A conceptual graph is a useful tool for visualizing and understanding the interconnections between different thoughts in relation to a certain topic. This would not only help the students better understand the material, but it would also encourage them to think critically and ask questions. These days, content-driven education focuses on rote memorization rather than helping pupils build their critical thinking abilities.

Topics such as the nature of concepts and their relationships with one another will be thoroughly covered in this chapter as we delve into the subject of Conceptual Graphs. The mathematical foundations of CGs are presented in a nutshell, and the steps to define a concept are shown with many instances of concepts. To paraphrase John Sowa, "to express meaning in a way that is logically exact, human accessible, and computationally tractable" is the goal of conceptual graphs.

In mathematics, a bipartite directed finite network is called a concept graph (CG). The graph's nodes stand for ideas or relationships. The concept node represents the ideas of entities, attributes, states, and events, whereas the relation node shows the relationships between the concepts. Two values—a type and a referent or marker—may be associated with a node, which can be a notion or a relation. A referent may be either a person or the singular generic referent. For this reason, CGs consist of sets of both concept types and relation types. The original discovery of CGs was attributed to John Sowa's work on database interfaces (Sowa 1976). It offered visual depictions of CGs that could be used to connect queries expressed in plain language with conceptual schemas. Declarative component graphs (CGs) with actor nodes representing database relations or functions were an integral part of each model. The model defines convolutional gradients and how they are used to represent natural language texts. Entities, attributes, events, and actions are all represented by concept nodes, while the nature of the link between two concept nodes is expressed by relation nodes.

2.12 Linguistic and Psychological Perspective

The mental model hypothesis states that people's understanding of the world is based on the mental models they create for themselves. This has led to the revelation of many fundamental issues in several branches of cognitive science. The most important questions are answered in the following list associated to the field of psychology: a) The following problems need answers: how the brain stores mental models; how these models interact with perception, memory, and learning; and how behavior influences or is impacted by these models. b) Linguistics is the field that this investigation focuses on, and the issue at hand is the relationship between a word, its meaning, and a mental representation of that object. How do models relate to the regulations that control syntax and semantics? c) Philosophy: How do our internal models of the world relate to our understanding of it? How do models fit into the reasoning process, and how does informal reasoning relate to formal logic? d) CS: How may one's mental picture of the world be codified into a digital format? Computer science is the domain in which this inquiry lies. To describe these models and build linkages to other systems, which programming languages and tools are necessary? Do the models have what it takes to back an interface that people would say is easy to use? From a linguistic and psychological perspective, the main points are how people understand and store ideas in their brains, and how these mental representations of concepts are conveyed *via* our language.

Also, ethical considerations of the human thought process is necessary. A framework for ethics in AI is referred to in ref. [15].

The question that needs answering is whether the words themselves convey ideas or if they are just symbols for the notions that humans understand and model in their brains. To perceive something is to build a working model that both represents and interprets input from the senses. Perception, the act of constructing a functional model, has two parts: the Conceptual Graph and the Percepts. One way to depict the interconnections between various perceptions is *via* a conceptual graph. A static representation of the data obtained by the senses is called a percept. Each perception serves a certain purpose, which is defined by the conceptual relations. When the letter e represents an external object or scene, the perceptual process generates a structure called a conceptual graph. Boxes in diagrams stand in for ideas, circles for conceptual links, and arrows connecting boxes to circles indicate arcs. Square brackets, like [ARCH], and rounded parentheses, like (), may be used to shorten the boxes in a linearly

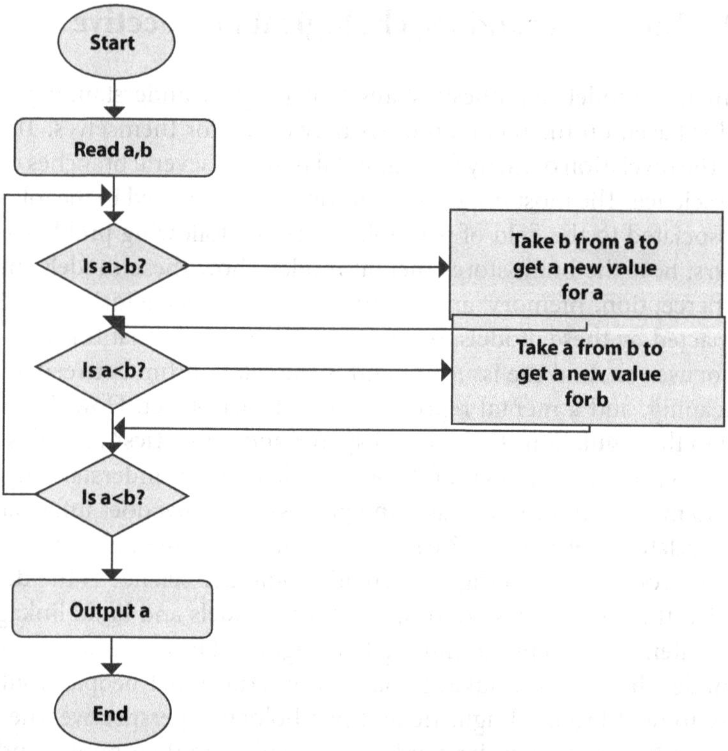

Figure 2.5 Flowchart of proposed work.

organized text (PART). Figure 2.5 displays a flowchart for the proposed conceptual reasoning logic.

There is no limit to the number of arcs that conceptual connections may have; nonetheless, the vast majority of them are dyadic. Monadic constructions include some, such as the marker for the past tense (PAST) or the negation (NEG). For the example sentence "a gap lies between a brick and a brick," the conceptual graph will look like the graphic that can be seen further down in this article. The conceptual relation (BETW) may be represented as a triangle.

2.13 Image Synthesis from Layouts

Scene layouts are the intermediary phases used for making images from scene graphs. The need to manually construct scene graphs was, however, removed by providing an explicit framework for instantaneously creating images from layouts. The object classes and bounding boxes are established first. The next step is to use the coarse layouts discussed before to create a diverse set of images. To build on their earlier suggestions, they clarified the loss functions and added object-wise attention to their proposed framework, which allowed them to enlarge the object feature map module. Similar work is discussed in recent literature on scene graph image synthesis. Figure 2.6 shows the architecture of the classification network based on a generative adversarial network (GAN).

In contrast, their method enhances layout realism *via* the use of scene graph-based retrieval. Consequently, their method is a hybrid of procedures for creating pictures from scene graphs and layouts. They unveiled an Object-Centric Generative Adversarial Network (OC-GAN) that uses

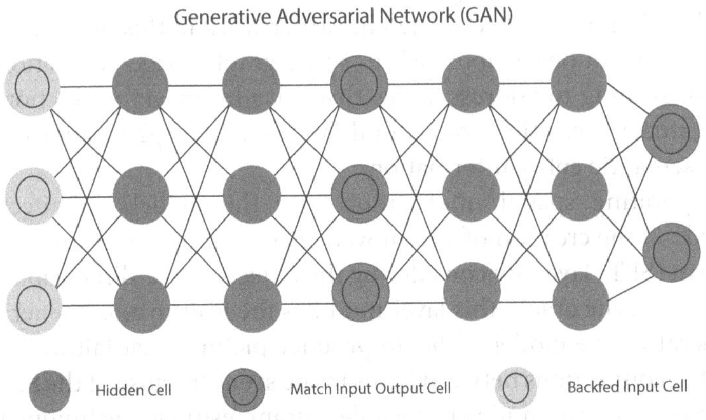

Figure 2.6 Classification network architecture.

the scene graph similarity module to understand the objects' spatial representations in a scene's layout. From a distance, the images generated by most SL2I generation algorithms look to be realistically reproducing the input layouts, which is one of their method's drawbacks. A closer look at these photos, nevertheless, reveals that neither context awareness nor location sensitivity are there. One piece of work that has been completed so far has used a context-aware feature transformation module to overcome these constraints. Their proposed method ensures that all items' generated features are regularly updated. Simultaneously, for feature maps, the Gram matrix is calculated to record the inter-feature correlations that account for the objects' position sensitivity.

2.14 Method Comparison

The methods used in this research, which were based on SG2I and SL2I, will be detailed later on. Based on the revolutionary work of the SG2I methodology, we landed on four separate ways that share the same input and training data.

A strong grasp of computer vision and computer graphics is required to create complicated scene pictures from actual objects. Making a photorealistic picture from the items and their interactions represented in a scene graph is the main objective of scene graph image production. One possible input and output for this is a scene graph.

It takes a lot of work to create a scene graph-based mage, which involves building recognized pieces in complicated scenes. The vast majority of picture-generation synthesizing scene graph algorithms depend on GCN. The feasibility of developing a picture retrieval system based on a scene graph description was investigated in earlier work in this field. Eventually, sg2im—a revolutionary approach to scene graph image generation—was unveiled. Aiming to address issues that arise in image production when using textual descriptions or natural language, the sg2im approach pertains to semantic entity information.

Incorporating spatial information into the model's architecture is achieved *via* the creation of an innovative layer known as a spatial feature transform (SFT) layer. By considering the objects' three-dimensional locations in the scene graph, this layer modifies the feature maps generated by the generator. The model is able to produce pictures that faithfully depict the spatial connections between items in the scene because of the SFT layer.

The first one was an idea to include a graph estimate technique for the classification issue into a deep architecture with a manageable learning complexity. Propagation models and traditional CNNs are used in the construction of both the first and second models. It is a technique the manufacturing process using an external item crop as an anchor. Their suggested strategy deviates from existing methods in three significant ways. The process begins with using object crops externally, continues with converting layout masks into pictures using a Crop Refining Network, and culminates with the introduction of a Crop Selector to automatically determine the optimal crops from the object database.

There was talk of an interactive usage of recurrent neural networks to create pictures from scene graphs, with the ability to edit cumulative images and keep the original image content intact. The procedure consists of three distinct phases, with more difficult tasks introduced at each level. At each stage, the scene graph is expanded to include new nodes and edges, giving the GCN more data for layout development. For the purpose of making bounding box predictions, they used a Scene Layout Network (SLN), a layout-creating network. Atop the architecture that was set up in, this network is used. Unsupervised image generation is made possible by their suggested technique, which combines the GCN with the adversarial image translation method. The generated images still have room for development; for example, the details are hazy, and the created items do not correspond to the given scene graphs. Using a scene context network, another piece of work further improved upon it. Using a context-aware loss and two additional metrics to assess the degree to which generated images conform to scene graphs, this research aimed to enhance picture matching. These metrics are the relation score and the mean opinion relation score. The degree to which the generated images correspond to the scene graph is evaluated by these two metrics.

The GCN could sometimes become confused about the relationships between attributes, and discovering the correct connection can be a tedious procedure; both of these issues are problems with GCN-based techniques. A canonical representation based on an approach to visualizing scene graphs was suggested as a means to achieve this goal; this representation respects the relationships between characteristics as it maintains the graph's information while canonicalizing.

2.15 Conclusion

To conclude this, this is a giant leap for computer vision to create a scene graph from static images. By identifying objects coupled in their reciprocal relatedness, it identifies the possibility of more structured marketing visionary of instances. This process includes object recognition and connection modeling that also helps us to understand visual material more easily and also is used in many different fields.

Graph Convolutional Networks (GCNs) and other graph-based methods, as well as message-passing networks, have considerably boosted the detection of relations between objects in images. These are methods that have been developed through deep learning now allowing for an improvement in the precision and depth of scene interpretation.

Furthermore, the scene graph construction could find its applications in other domains such as visual question answering, picture captioning, recommendation systems, and robotics [16]. All those created scene graphs better inform intelligent decision making, which could help robots map out pathways, make recons, and field more complex image-related questions.

Here are some known issues for high-level reading: It cannot scale to unlabeled data. It still needs annotations on the final dataset, which can be much improved but is not perfect Interpreting the outputs of these models is still an area of active research. However, these problems are still around because of continued research in graph-based learning and computer vision (where the latter is feeding into the former a bit reductively), to form scene graphs that are closer and closer approximations to large-scale, unified visual understanding.

In conclusion, compounding scene graphs from photos opens an entirely fresh area of understanding visual information that may have implications through computer vision in hundreds of regions and fields of artificial intelligence.

References

1. Anderson, P., He, X., Buehler, C., Teney, D., Johnson, M., Gould, S., Zhang, L., Bottom-Up and Top-Down Attention for Image Captioning and Visual Question Answering. *Proceedings of the IEEE Conference on Computer Vision and Pattern Recognition (CVPR)*, 2018.
2. Zellers, R., Yatskar, M., Thomson, S., Choi, Y., Neural Motifs: Scene Graph Parsing with Global Context. *Proceedings of the European Conference on Computer Vision (ECCV)*, 2018.
3. Xu, D., Zhu, Y., Choy, C.B., Fei-Fei, L., Scene Graph Generation by Iterative Message Passing. *Proceedings of the IEEE Conference on Computer Vision and Pattern Recognition (CVPR)*, 2017.

4. Li, Y., Ouyang, W., Zhou, B., Wang, X., Scene Graph Generation from Objects, Phrases and Region Captions. *Proceedings of the IEEE International Conference on Computer Vision (ICCV)*, 2017.
5. Lin, T.Y., Maire, M., Belongie, S., Hays, J., Perona, P., Ramanan, D., Dollár, P., Zitnick, C.L., Microsoft COCO: Common Objects in Context. *European Conference on Computer Vision (ECCV)*, 2014.
6. Johnson, J., Krishna, R., Stark, M., Li, L.J., Shamma, D.A., Bernstein, M., Fei-Fei, L., Image retrieval using scene graphs. *Proceedings of the IEEE Conference on Computer Vision and Pattern Recognition (CVPR)*, 2015.
7. Zhu, Y., Urtasun, R., Salakhutdinov, R., Fidler, S., Lin, D., Structured prediction models for semantic segmentation. *Proceedings of the IEEE Conference on Computer Vision and Pattern Recognition (CVPR)*, 2017.
8. Gupta, A. and Malik, J., Visual semantic role labeling: A task for structured perception. *Proceedings of the IEEE International Conference on Computer Vision (ICCV)*, 2015.
9. Krishna, R., Zhu, Y., Groth, O., Johnson, J., Hata, K., Kravitz, J., Fei-Fei, L., Visual Genome: Connecting language and vision using crowdsourced dense image annotations. *Int. J. Comput. Vision*, 123, 1, 32–73, 2017.
10. Johnson, J., Gupta, A., Fei-Fei, L., Image Generation from Scene Graphs. *Proceedings of the IEEE Conference on Computer Vision and Pattern Recognition (CVPR)*, 2018.
11. Dai, J., Qi, H., Xiong, Y., Li, Y., Zhang, G., Hu, H., Wei, Y., Deformable Convolutional Networks. *Proceedings of the IEEE International Conference on Computer Vision (ICCV)*, 2017.
12. Kipf, T.N. and Welling, M., Semi-Supervised Classification with Graph Convolutional Networks. *International Conference on Learning Representations (ICLR)*, 2017.
13. Huang, G., Liu, Z., Van Der Maaten, L., Weinberger, K.Q., Densely Connected Convolutional Networks. *Proceedings of the IEEE Conference on Computer Vision and Pattern Recognition (CVPR)*, 2017.
14. Basheer, S., Singh, K.U., Sharma, V., Bhatia, S., Pande, N., Kumar, A., A robust NIfTI image authentication framework to ensure reliable and safe diagnosis. *PeerJ Comput. Sci.*, 9, e1323, 2023.
15. Sharma, V., Mishra, N., Kukreja, V., Alkhayyat, A., Elngar, A.A., Framework for Evaluating Ethics in AI. *2023 International Conference on Innovative Data Communication Technologies and Application (ICIDCA)*, Uttarakhand, India, pp. 307–312, 2023, doi: 10.1109/ICIDCA56705.2023.10099747.
16. Borkar, K.K., Aljrees, T., Pandey, S.K., Kumar, A., Singh, M.K., Sinha, A., Singh, K.U., Sharma, V., Stability Analysis and Navigational Techniques of Wheeled Mobile Robot: A Review. *Processes*, 11, 3302, 2023. https://doi.org/10.3390/pr11123302.

3

Transformation from CNN to Graph-Structured Data: Node Classification and Edge Prediction

R. Vijayarangan[1]*, R. Satish Kumar[2], K. Umadevi[3] and K. Ashok Kumar[4]

[1]*Sengunthar Engineering College, Tiruhengode, India*
[2]*Department of Computer Science and Engineering, Sengunthar College of Engineering, Tiruchengode, India*
[3]*Department of EEE, Sengunthar Engineering College, Tiruhengode, India*
[4]*Department of CSE, Sengunthar Engineering College, Tiruhengode, India*

Abstract
The proposed study investigates how to structuralize convolutional neural networks (CNNs) for better inference of graph-structured data and refine both node classification and link prediction matters. This approach leverages graph representations to exploit the underlying relational information among data points. CNN architecture can work with graph structures, and they call it a "Globally-Interactive Locally-Propagated" (GILP) CNNs for node-level classification. We introducc a technique that seeks to generalize convolutional neural networks (CNNs) from grids and images to arbitrary graphs, while also learning tools for feature transformation in the context of prediction on sets of nodes or edges based on attributes of these. Our results from empirical experiments and comparative analysis have shown that this approach can outperform traditional vector mapping methods. The results underscore the promise of marrying CNNs and graph-based frameworks to improve representation learning and predictive modeling in graph-structured data spaces.

Keywords: CNN (convolutional neural network) transformation, graph representation, node classification, edge prediction, graph-structured data, relational information integration

**Corresponding author*: phdcsevijayarangan@gmail.com

Malini Alagarsamy, Rajesh Kumar Dhanaraj, J. Felicia Lilian, Vandana Sharma and Gheorghita Ghinea (eds.) Graph Convolutional Neural Networks for Computer Vision, (47–72) © 2026 Scrivener Publishing LLC

3.1 Why Graphs

Because of their exceptional capacity for expressiveness, graphs are among the data structures that are the most flexible. Machine learning models have been effectively implemented in a range of domains [1], with the goals of extracting and predicting information on data resting on graphs, as well as modeling complicated components and their interactions.

The number of schools and colleges that can be attended in India is much higher. However, many students stop participating in their academic pursuits while they are still enrolled in school for a variety of reasons. The primary causes are the current state of the economy, a decreased level of self-interest, concerns pertaining to women, insufficient higher educational institutions, and so on. It is beneficial to the lives of the student to have an early prediction of what their future job will be like. Data analysis techniques [2] have traditionally been used to make predictions about a student's future career. In this stage, the data are gathered from the educational organizations and then manually examined. It is a procedure that requires a lot of time. As a result of the development of various communication methods, computer algorithms are now being used to forecast a student's future profession. When it comes to estimating the student's future career path, machine learning (ML) [3] is one of the key methodologies that plays a vital role. The sections that follow provide examples of some of the most significant methodologies used in the field of career prediction.

According to ref. [4], the majority of firms kept the personnel database up to date. The vast majority of the personnel database comprises employee selection history, particulars about the recruiting process, information regarding the employer's involvement, and employee biodata. One of the approaches that is used to obtain information from a massive amount of data is known as data analysis. The use of ML strategies is becoming more relevant in our day-to-day lives. It has been put to use in the detection of back frauds, the recommendation of promotions for e-commerce, the control of driverless automobiles, and the prediction of illnesses. To accurately predict a student's employability once they have graduated from an advanced learning school, ML is an essential technology that cannot be avoided. Figure 3.1, which is seen below, presents an overview of the process that the ML approaches follow.

The students' employability abilities are evaluated using a variety of criteria, including their main subject, grade point average, extracurricular activities, and internship experience. The characteristics of employable UG (undergraduate) students are broken down in Figure 3.2, which can be found below.

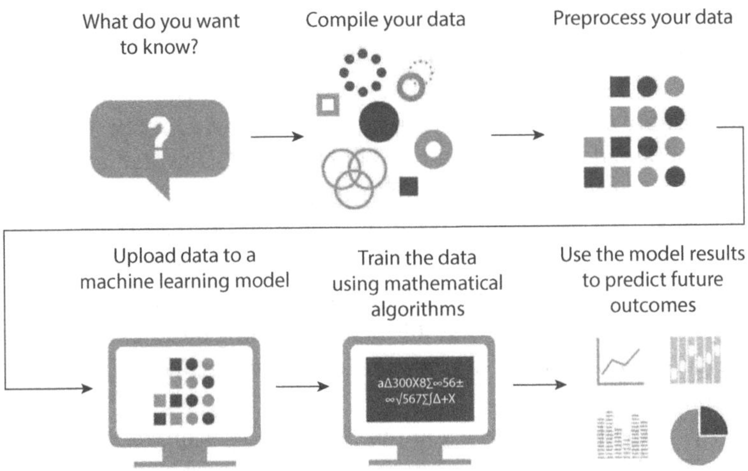

Figure 3.1 Outline of ML techniques.

- Predicting protein interactions for drug research, projecting traffic patterns on road networks, and recommender systems based on data from social networks.

One kind of graph neural network, known as graph convolutional networks (GCNs), makes use of convolutional aggregations (GNNs). Conventional convolutional neural network (CNN) architectures have shown to be quite successful in machine learning, especially when it comes to solving computer vision problems. Nevertheless, when dealing with a collection of problems that have a graph structure, it becomes challenging to express them using traditional neural networks. Evidence suggests that

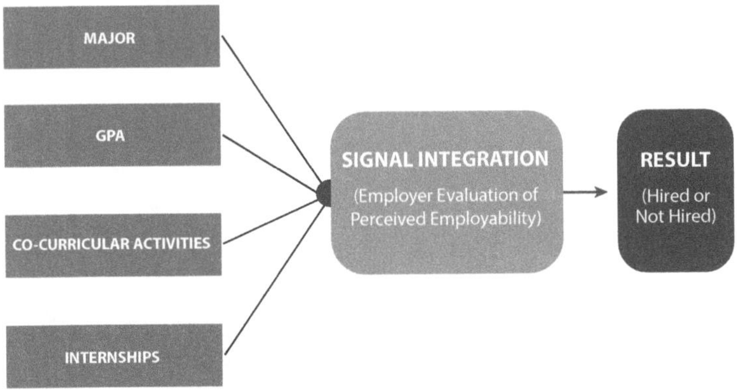

Figure 3.2 Major employability features for the UG students.

GCNs may provide comparable or better performance when applied to certain computer vision problems. As was indicated in the last essay, GCNs have been used to discover answers to several problems. Applications, such as visual question answering and picture categorization, are a few examples [5].

The generalized convolutional neural networks (GCNs) fall into two main types as follows: spectral and spatial. To summarize, GCNs may be either spectral or spatial. The former is defined in the data's spectral domain using the graph Fourier transformation, while the latter is an aggregate of the node representations of its neighborhood in the original graph domain. On the data's spectrum domain, spectral GCNs [6] are defined using the graph Fourier transform. Figure 3.3 illustrates real-world machine learning applications mapped on a graph.

Graphs provide a versatile toolbox for solving a broad range of challenges in machine learning.

Assigning a given graph to one of many predefined classes is the goal of graph classification.

- Node Classification: Identify the remaining node classes in a network when node labels are incomplete.
- Predicting Links: Given a graph with an unfinished adjacency matrix, find out whether every pair of nodes is linked.
- Graph Clustering, or Community Detection, is dividing a network into groups of nodes according to the edges in the graph.

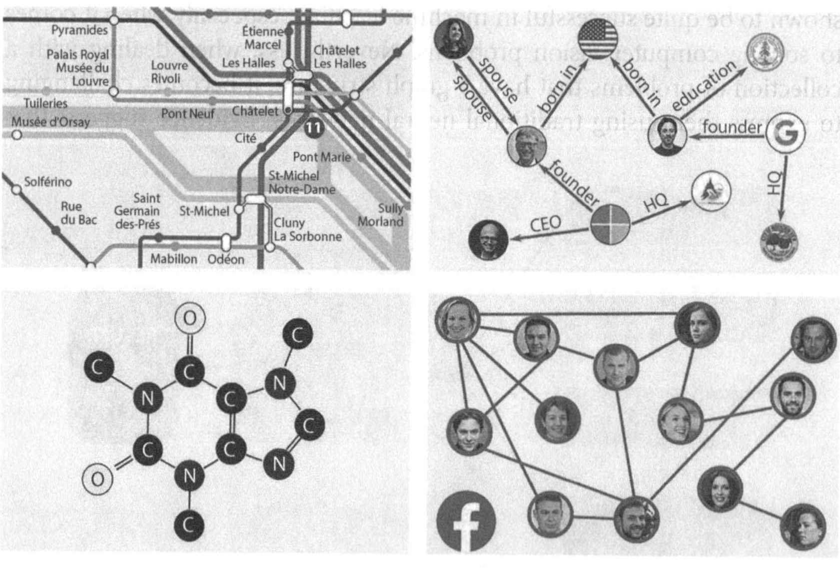

Figure 3.3 Applications of machine learning on graphs (image by author).

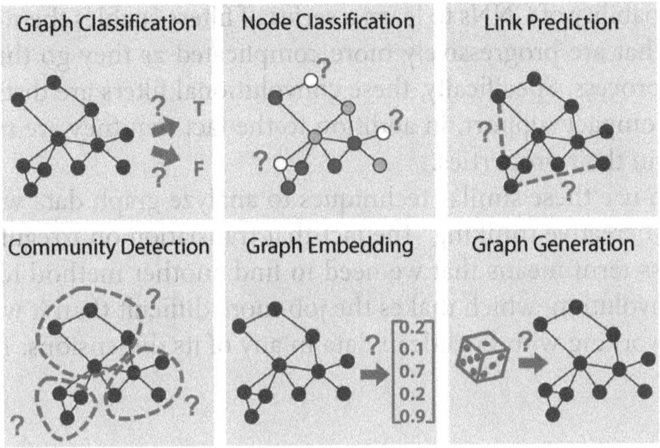

Figure 3.4 Machine learning tasks on graphs (image by author).

When given a graph, graph embedding converts it to a vector while preserving important details. Graph generation is a process that involves learning a distribution from a set of provided graphs and then using that distribution to generate new graphs that are similar to the original set. Figure 3.4 outlines various machine learning tasks applicable to graph data.

The lack of structure and conformity to Euclidean geometry in graph data makes it difficult to intuitively build machine learning models to address these issues. Finding a way to analyze the data conveyed by connections between nodes is not simple, but it is crucial since these data are essential for the network to function.

Here, we will take a look at how graph convolutional networks (GCN), a variant of classic convolutional neural networks (CNN) tailored to graph-structured data, may solve the problem. This article mostly relies on the study conducted by the authors of ref. [7].

It has been shown that convolutional neural networks (CNNs) are very good at extracting complex data, and convolutional layers are now the foundation of many deep learning models. CNNs have shown their efficacy with data that have several dimensions, such as the following:

- Analyzing auditory signals in one dimension—for instance, to classify sounds
- In two dimensions, for picture analysis (e.g., early tooth decay detection)

Using three dimensions, images may be processed, such as in magnetic resonance imaging (MRI) brain registration.

The capability of CNNs to learn a series of filters enables them to extract patterns that are progressively more complicated as they go through the learning process. Specifically, these convolutional filters are distinguished by their compact support, in addition to the fact that they are translation invariant in their properties.

We can use these similar techniques to analyze graph data with only a little bit of creative thinking. The fact that translation on irregular graphs is a useless term means that we need to find another method to describe graph convolution, which makes the job more difficult than it would be if we were working with Euclidean data in any of its dimensions: 1D, 2D, or 3D.

3.1.1 The Definition of "Graph Convolution"

Taking the product of the translated functions defines convolution on Euclidean domains. Nevertheless, we will need to take a different tack to define translation on irregular graphs.

Here, the key idea is to use a Fourier transform [8]. In the frequency domain, the well-defined convolution of two signals is obtained by taking the component-wise product of their transformations, according to the Convolution Theorem. This happens due to the lack of clarity in the convolution of the two signals. We might define convolution as the product of the Fourier transform and the function if we knew how to compute the Fourier transform of a function that was defined on a graph.

$$x * y = F^{-1}\{F\{x\} \cdot F\{y\}\} \tag{3.1}$$

Our approach will be based on the classical Fourier transform, which will help us solve this issue. I will use the example of a real-line-defined function. The function is projected onto an orthonormal basis of sinusoidal waves to yield its breakdown in frequency terms, which is known as its Fourier transform. Surprisingly, the eigenfunctions of the Laplacian are these following waves:

$$\frac{d^2}{dx^2}u = \lambda u \rightarrow u_\omega(t) = e^{i\omega t}\ \hat{x}(\omega) = \int_{-\infty}^{+\infty} \Sigma e^{-i\omega t} x(t) dt \omega \in R \tag{3.2}$$

Expanding on this concept, we may say that a function's Fourier transform is its projection onto an orthonormal basis of Laplacian eigenfunctions as follows:

$$Lu = \lambda u \rightarrow \{u_\lambda\} orthnorm.basis\, \hat{x}(\lambda) =< u_\lambda, x\rangle \tag{3.3}$$

The formula for the Laplacian matrix in graph theory is L = D – A, where D is the degree matrix that contains the number of edges linked to each vertex, and A is the adjacency matrix that shows whether two vertices are connected by an edge for each pair of vertices. To rephrase, the Laplacian matrix (L) is defined as D minus A, where D is the diagonal matrix that contains the number of edges connected to each vertex (the degree matrix).

The Laplacian L will be a real symmetric positive semi-definite matrix if we assume that the edges of the graph do not point in any particular direction (however, this definition may be extended). As a result, it may be characterized as the projection of a graph signal onto an orthonormal matrix U, which diagonalizes it. This signal has the form of an N-dimensional vector [9] with values at each of the N vertices as follows:

$$\begin{aligned} Lu = \lambda u \rightarrow U = [u_1, \ldots, u_N] \,\|\|\, \hat{x}_l = u_l^T x l \in \{1, \ldots, N\}\; s.t.L \\ = U^T \Lambda U \,\|\|\!\rightarrow \hat{x} = U^T x \end{aligned} \tag{3.4}$$

3.1.2 Graph Fourier Transform

A picture is worth a thousand words; therefore, let us look at some examples to understand what this means. A regular two-dimensional grid's Delauney triangulation graph reveals that the vibration modes of a free square membrane completely match the Fourier basis of that graph. The eigenfunctions [10] of the Laplacian are the fundamental modes of a vibrating plate; therefore, this makes sense. Figure 3.5 shows the square root of the Laplacian's eigenfunctions in graphical form.

It is still possible to see the graph's vibration modes in a randomly created graph by examining its orthonormal Fourier basis. Figure 3.6 depicts Laplacian Eigenfunctions for an arbitrary graph.

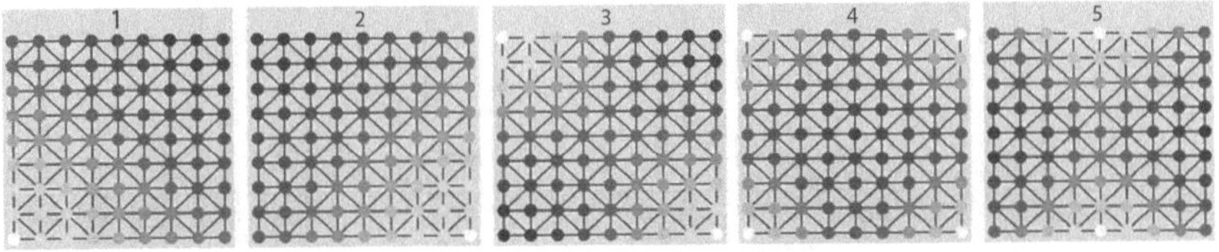

Figure 3.5 The square root of the Laplacian's Eigenfunction (image by author).

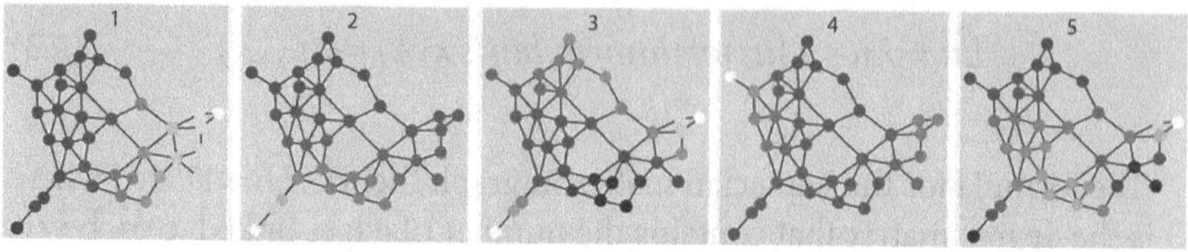

Figure 3.6 The Laplacian's Eigenfunction for an arbitrary graph (image by author).

The structure of a graph convolutional network can now be understood thanks to our familiarity with the definitions of graph convolution and graph Fourier transform.

3.1.3 Developing a Comprehensive Neural Network

All convolutional networks used for picture identification have a similar structure in their design. This holds true for both basic and advanced neural networks, such as ResNet and VGG16.

1. The H × W × C input picture is processed *via* a sequence of pooling layers and localized convolution filters to extract features.

   Using a global pooling layer or another method, the feature channels that are produced are then transformed into a vector of given size.

2. The last step in producing the categorization result is by using a small number of fully connected layers [11].

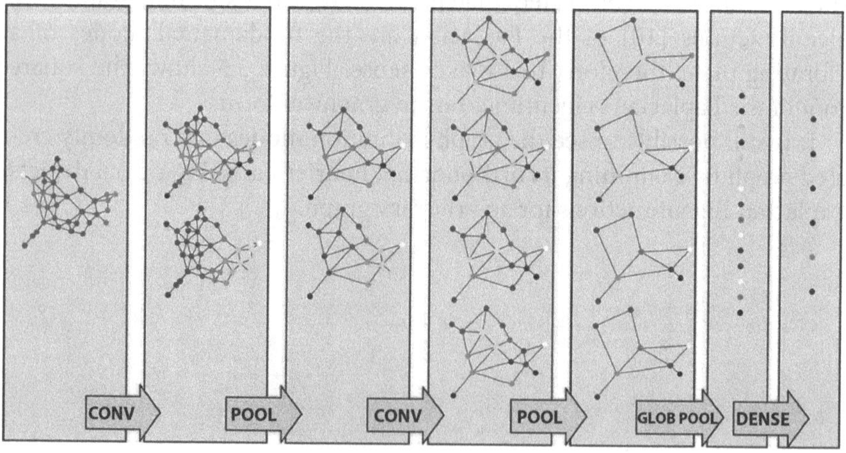

Figure 3.7 Graph convolutional network (image by author).

A graph convolution network's design is structurally identical to any other. Figure 3.7 visualizes the architecture of a graph convolutional network.

- The following components constitute our input for a GCN:
- x is an N × C array that contains, for every N node in the graphs, the N × N adjacency matrix.

We still need to construct pooling layers and assure that convolution filters have compact support before we can completely grasp the architecture illustrated above.

Any graph clustering technique that can join groups of nodes together while keeping local geometric structures intact may be used for pooling layers. A rapid greedy approximation is utilized in practice since optimum network clustering is an NP-hard issue. The GRACLUS multilevel clustering technique is a favored option [12].

How can we ensure that our GCN's convolutional layers function locally in relation to the convolutional filters' compact support? In most cases, g acts as a filter for an input x.

$$x * g = U\,\hat{g}(\Lambda)U^T x = \hat{g}(L)x \tag{3.5}$$

Nevertheless, such a filter lacks compact support in the absence of additional assumptions, and learning all the components of (Λ) is O(N) complicated. To resolve these problems, we shall substitute a polynomial parametrization of degree K as follows:

$$\hat{g}_\theta(L) = \sum_{k=0}^{K-1} \Sigma\theta_k L^k \tag{3.6}$$

The learning difficulty is reduced to O(K) since we just need to learn θ 0,..., θ {K−1}. Moreover, it can be shown that it is K-localized, meaning that the value at node j of Ę centered at node I is zero if there are more than K minimal edges connecting any two nodes on the network. The most common value is K = 1, which represents 3 × 3 convolutional filters applied to pictures. Figure 3.8 compares compact filtering on regular and irregular structures.

One last thing about optimization in computational models. Due to the multiplication involving U, the computational cost of the filtered signal (L)x remains as high as O(N^2). Recasting the polynomial as a Chebyshev

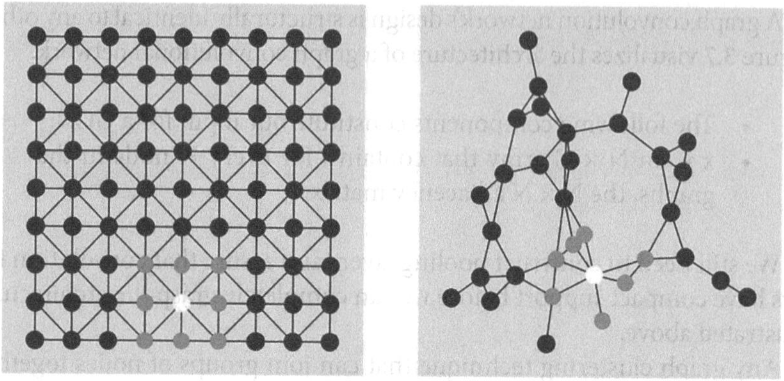

Figure 3.8 Compact filtering on a square and on an irregular (image by author).

polynomial, which has a very practical recursive formulation, may bring the cost down to O(EK) (where E is the number of edges) [13].

The authors of ref. [14] say that DT is the most familiar, simple, and easiest ML technique offering the best solution for classification issues. DT is the basic algorithm; from DT, various advanced techniques are derived. The common advanced approaches are bagging, RF, and boosting. The familiar DTs are C5, ID3, and CART 4.5. Commonly, in DT, a single node represents an input identifier (X) and fragments that identifier. If the identifier is a number, the leaf of the DT is known as a terminal-type node, and the output identifier (y) is important for the forecasting. Commonly, DT follows to choose the root-type node. Then, measure the IG value or entropy value of every node before dividing the node. Choose the node that contains a high IG value and low entropy value. Then, divide the node and repeat the same process. The previous step is repeated until there is no option to divide or the entropy value is the lowest. Entropy is the best scale to identify the uncertainty level or data randomness. If scale is used to identify the amount of entropy, the value is decreased to the previous level after division [15].

$$H(s) = \sum -p(x)log_2 p(x) \quad (3.7)$$

$$IG(A,S) = H(S) - \sum p(t)H(t) \quad (3.8)$$

According to ref. [16], DT is one of the supervised-type algorithms. Here, the tree-type structure is created based on the entropy value.

The high-depth tree produces results with better accuracy. It is very helpful to predict the student's performance from the current dataset. DT pruning process or branch cropping approaches are avoiding the overfitting issue. The pruning process is easily implemented by managing the attributes of API sklearn [17].

Merits of Decision Trees

Decision trees are easy to visualize, understand, and interpret.

- DTs indirectly do the feature or identifier selection process.
- It manages numerical, categorical, and multi-output issues.
- DT requires less amount of effort from the user side for preparing data.
- Nonlinear-type associations among the arguments do not influence the performance of the tree.

Demerits of Decision Tree

- Learners of the DT can make a complex tree if they do not generalize the information in a well-defined manner. This is called overfitting.
- DTs may be unbalanced because the minor change in the given data affects the entire tree structure. This is known as a variance.
- The greedy approach does not issue an assurance to provide the optimal DT [18].
- DT learners make biased-type trees if any classes are dominated.

3.2 SVM (Support Vector Machine)

SVM is one of the supervised-type ML algorithms, which is commonly offered as a solution for classification- and regression-type issues. It follows the standard process. In this algorithm, each data part is placed in the nth dimensional region. Here, n denotes the total number of attributes, and every attribute has the value of the concerned coordinate. Then, it categorizes by receiving a separate hyperplane that divides the classes [19].

Roshani Ade *et al.* (2014) say that the SVM training process needs the description of an exact quadratic-type programming problem. Memory space is required to train the concerned dataset by SVM, and it permits the management of a huge volume of data. Normally, SVM changes the missing contents and the nominal-type identifiers into binary types. The training process may be executed with the help of polynomial- or RBF-type kernels. Pairwise-type classification is used to solve multiclass-based issues [20].

According to the study in refs. [21] and [22], the SVM technique provides major benefits than other ML approaches. Due to that reason, use the SVM technique to measure the pass rate of the student and the accuracy value of the particular model. The SVM approach is a nonlinear-type method. It is inappropriate for the sample dimension, and the smaller number of samples also maintains durable flexibility for the derived sample [22]. The actual statement of the normal SVM technique is decreased to the subsequent problem.

From Equations (3.7) and (3.8), we derive $\xi_i >= 0, x_i \in R^m$ is denoted as the input vector of the SVM, $y_i \in \{-1, 1\}$ is the group of the value x_i, $i = 1, \ldots\ldots, w$ describes the hyperplane usual vector, b denotes the offset of the hyperplane, ξ_i represents the relaxation-type variable, and penalty factor is represented as the variable C. Specifically, dependency of the features is specified based on the coefficient value, with the help of the grid finding technique to find the common kind of kernel methods. The kernel-type method is described as follows:

In SVM, the penalty factor C and the kernel method are not suitable, which leads to the overfitting issue. The grid selection technique was offered to solve this issue and argument optimization. Testing the accuracy value used a 10-fold cross-validation function.

3.3 XGBOOST

XGBOOST is one of the common ML algorithms that executes better on most of the classification processes. The authors of ref. [23] apply the XGBOOST method for forecasting the postgraduation student's plan. It is a kind of DT technique and creates a model using phase repetitive ways [24].

Chen T *et al.* (2016) describes the XGBOOST model $\varnothing(x_i)$ in the following manner [25]. Here, the first term represents the conventional-type loss method, and the second term denotes the regularization-type component to overcome the overfitting issue. α represents the hyperparameter to decrease the capacity of the model. The Ω (f) value of every tree may be measured as follows:

$$\Omega(f) = \gamma^T + 1\lambda''m''^2 \quad (3.9)$$

Here, the variable T denotes the total number of leaves available in the tree, ω denotes the score value of all leaves, and the trade-off arguments are denoted by γ and $\boldsymbol{\lambda}$.

3.4 Artificial Neural Network (ANN)

ANN is one of the computational systems that depend on the framework of biological NN. In career prediction, the input layer of the ANN contains various academic parameters like percentage of the mark, co-curricular and extra activities, personal options, etc. Nearly 15 input parameters are used in the ANN input layer. Twenty input nodes are used, and the activation method passes the output value to the next level layer. The ReLu method is applied here instead of tanh and sigmoid function for better results. In career prediction, L1 regularizer is offered instead of L2 because it decreases the fewer significant attribute coefficient value to zero. The second-level layer consists of 20 nodes; it accepts the input value from the initial level layer. The final layer contains six output values that forecast

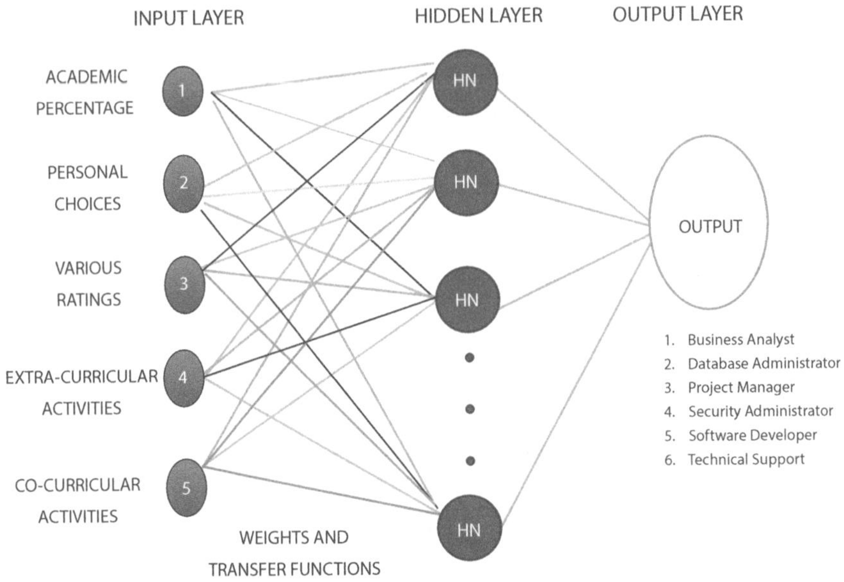

Figure 3.9 ANN framework for career prediction.

the domain, and it uses the multiclass-type classification in the final layer. The softmax method is applied in this model for classification, and it produces a better outcome. Figure 3.9 presents the ANN framework for career prediction.

Selecting the particular domain of the model fixes the possibility of every class, so it chooses the class with the highest possibility using the value of categorical-type cross-entropy value. The following Figure 3.10 describes the flow diagram of the ANN model used in the career prediction model.

To measure the bias value and the weight value, a feedforward approach is applied. The backpropagation technique is applied for updating the concern weights based on their layers. ReLu activation method is offered due to its features that provide exact results and a less costly process. This is described by the following function:

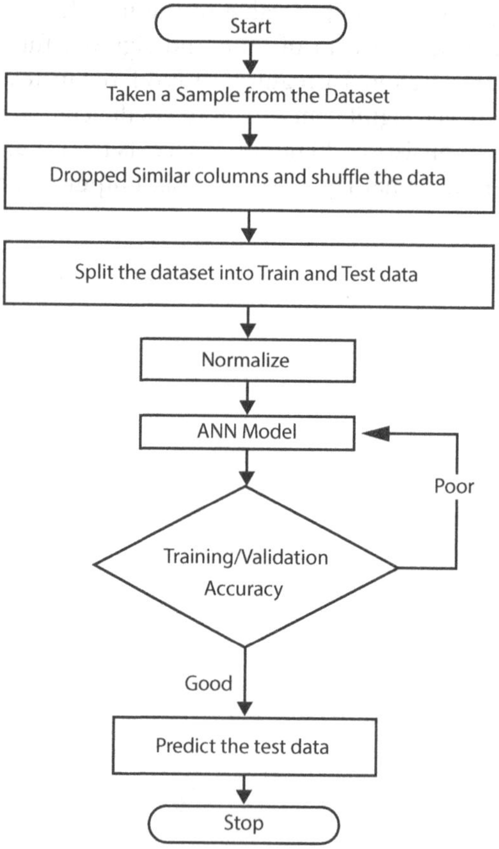

Figure 3.10 ANN process flow in career prediction.

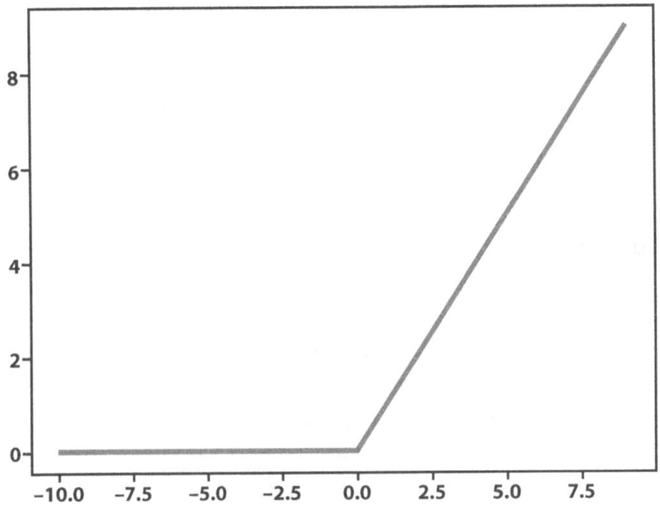

Figure 3.11 Graphical description of ReLu.

$$g(z) = \max(0, z) \quad (3.10)$$

The pictorial illustration of the ReLu method is shown in Figure 3.11.

The abovementioned model measures the loss value by the categorical cross-type entropy method. In this model, a loss is computed by the log value of $y\theta$ (predicated data) and the y (actual result). Then, this model offers Adam's optimization approach for renewing the biases values and the weights. For this type of computation of ($V\theta\,(x)$) variance and ($S\theta\,(x)$) SD of the weight, the value is measured using moment- and propagation-based updating. Initializing the values is done as follows:

$$Vd\theta(x) = 0, S\theta(x) = 0, Vdb(x) = 0, Sdb(x) = 0 \quad (3.11)$$

$$Vd\theta(x) = \beta_1 * Vd\theta(x) + (a - \beta_1) * d\theta(x) \quad (3.12)$$

$$Vdb(x) = \beta_1 * Vdb(x) + (1 + \beta_1) * db(x) \quad (3.13)$$

$$Sd\theta(x) = \beta_2 * Sd\theta(x) + (1 - \beta_2) * d\theta(x) \quad (3.14)$$

$$Sd\theta(\mathrm{x}) = \beta_2 * Sdb\theta(x) + (1 - \beta_2) * d\theta(x) \quad (3.15)$$

$$Sdb(x) = \beta_2 * Sdb(x) + (1 - \beta_2)db(x) \tag{3.16}$$

The weight value and the biases are updated using from the above equations.

3.5 Auto Encoder (AE)

Forecasting the students' academic preference is a common research domain in EDM, and it uses ML and DM approaches to discover information from the educational dataset. But measuring the student's performance is a difficult task because various factors are considered for identifying a student's career performance. Bo Guo *et al.* (2015) say that the associations among the factors are an identifier for forecasting the participant's performance in the critical nonlinear method. Conventional DM and ML approaches do not support this type of data and issues. Bo Guo *et al.* (2015) designed a model to forecast the performance of the student by DL methods, which learn many layers of the representation automatically. Unseen layer features are pre-trained on a layer basis with the help of an autoencoder model, and the attributes are fine-tuned by supervised methods. The proposed model is trained using a real-time student database [16].

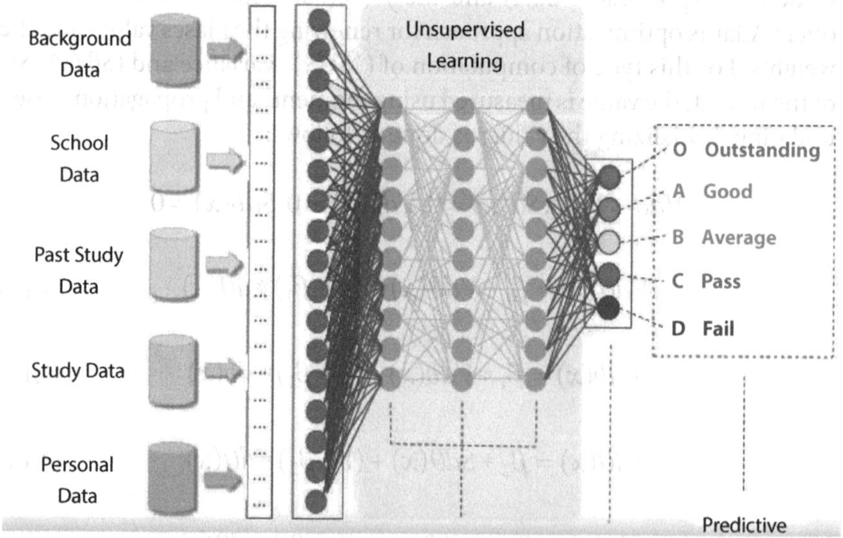

Figure 3.12 Overview of student performance forecasting system (Bo Guo *et al.*, 2015).

EDM is a new research domain. This approach is a combination of education, DM, and ML techniques. It uses ML and DM approaches to identify the factors from the educational data to discover the patterns and segregate the behavior of the students. The main aim of the EDM is to increase the educational-based output values and to attain improvements in the educational sector. The following Figure 3.12 describes the overview of the forecasting system used for students' academic performance.

Learning the multilevel association by unsupervised hidden-level layers is then followed by the backpropagation method, which is offered for fine-tuning the NN. The input value of this model is a flat vector value of the various types of student data, and the output data consist of many categorizations of softmax representing the pupil end exam score value (Bo Guo *et al.*, 2015). Training data contain different types of complex information.

3.6 Demographic and Related Data: Health Condition, Type of Gender, Age, Family Condition

Past Study Data: Entrance Mark, Primary-Level GPA

School Evaluation Information: Type of school, ranking of school study-related data: Each course score in high school and junior level.
Personal Information: the personality of the student, attention level, psychology-connected data.

Collected information is converted into numerical-type values and scaled and normalized by subtraction the mean value and dividing by the standard deviation value of the concerned elements to ensure that data falls within a similar range. Every input vector value is normalized; then, the whole dataset is balanced to reduce the redundant input values.

Student dataset consists of various types of information. Pre-processing tasks are applied to smoothen the given dataset. The flattened vector is denoted as v_1. Unsupervised-type techniques are offered to find the attributes from unlabeled-type data. Train the encoder for K unseen-type nodes in the unseen layer l with the help of the backpropagation technique to decrease the squared error between all m number of examples with a term penalty that constrains the number of units to achieve a less mean activation of the output value.

A ReLU method is as follows:

$$f(a) = \max(0, a) \tag{3.17}$$

The unsupervised method is applied in all unseen layers to declare the entire network systems' weight value W to the suitable values.

The entire deep system is pretrained and fine-tuned with the help of back-propagation. The final layer l_0 contains five neuron-type units; every unit is equivalent to the final exam-grade value. The dropped neurons are not activating the backpropagation and forward pass. Each time, the NN uses a different kind of framework but distributes similar weight values. Without the help of the above techniques, SPPN will be affected by the overfitting issue.

3.7 Naïve Bayes (NB)

NB is the simple method of Bayesian-type networks, as it assumes that all attributes are not based on the other attributes. The NB concept is normally used for batch-type learning. Due to the concept managing of every training data sample individually, it cannot do its process well. Based on the features of the incremental-type learning, the NB technique is trained with the following steps (Roshani Ade *et al.*, 2014):

a) Declare the identifiers' total and count equal to 0.
b) Read all samples; consider a single sample at a time.
c) Every $t\ (x, y)$ sample for training has its value related to it.
d) The count value is increased through the concerned samples used in training.
e) The possible values are computed by dividing the separate count value by training set samples of the same class features.
f) Measure the preceding possible values p(y) as the area of the whole training set in the y class [7].

3.7.1 K-Star

K-Star is a sample-type learner, and the testing samples are finalized using the label value if the samples depend on the similarity method. It uses an entropy-dependent distance method. The instance x possibility value is computed in class y by adding the possibilities from x samples to every sample that is an element of y. The possibility of every class is computed using Equation 3.18. The associative possibility received a guess of class circulation at the x sample (Roshani Ade *et al.*, 2014).

$$P*(C|x)\sum P*(x1|x) \tag{3.18}$$

NB is the most often used classification system because of the fact that it has the simplest form, the easiest comprehension, and the easiest interpretation. It requires less data training, can handle both category- and numerical-types of data, and can analyze a big number of data in a relatively short amount of time. A human choice may be thought of as being analogous to the ordered tree structure, which issues data about a classification to classify and segregate it within a class while also locating limitations in the most intelligible manner possible. Yuri Nieto and colleagues (2019) divide the collected data into the following two categories: graduated and not graduating. The most common use for the supervised-type learning method known as NB is in the realm of classification problems. Unexpectedly, DT can carry out operations on type-dependent identifiers that are continuous and categorical.

C5.0 is one of the several forms of NB, and it is the only one that has facilities for boosting and produces more accurate tree structures. In addition to this, it is possible to provide a weighting value to a variety of characteristics, it is possible to miscategorize types, and it automatically separates the data to aid in reducing noise level. In the process of developing NB, determining the value of IG, GR, GI, and entropy evaluation is very significant. The trend of learning is included into the split identifier selection process. The GI value is provided to facilitate the selection of the dividing identifier that is best appropriate for each node. The form that is utilized to express the split technique is as shown below.

The GI value may be quantitatively defined where J represents the number of classes (I 1,2,3,... J), and pi indicates the items that are associated with the group.

The collection of concerns has been partitioned into two distinct categories. By dividing, one may get the highest data of the lower part, the lowest data of the lower part, or the mean data of the two parts. Choose the last section of the split to ensure that the specified model is applied to nominal field values and only binary-type divisions. The pruning process is an additional method that may be used to bring the rate of incorrect classifications down. In addition, this method processes a high accurate value and reduces the processing time.

According to the research of Xiaofeng Ma *et al.* in 2018, in the field of education, the proportion of students who pass exams is used to characterize both the educational level of schools and the educational rate of the whole institution as a whole. The percentage of students that pass their degree program is determined by a variety of factors. In the field of education, it is crucial to determine the primary elements that have a significant impact on the percentage of students that pass the exam. In the NB methodology, the information value of entropy is provided as a means of determining the weight of the student identification. The NB framework is one of the branch frameworks, and it includes many different rule sets. In NB, the root node is responsible for the generation of the main attributes, while the leaf-type nodes represent the ultimate conclusion of classification. In DT, the construction of the branch building makes use of the method known as recursive-type partitioning. The if–then rule may be thought of as an illustration of the branch format for simply comprehending human beings. The dataset includes a wide variety of characteristics and traits. Classes are used to distinguish between distinct characteristics. The formula that is used to calculate the IE value for group D is Equation 3.19.

The IE value indicates the state of the node in the partition-type tree to locate the most appropriate identifiers for the dataset that has been provided. Choose the higher department value of the DT to tweak its arguments to provide a prediction about the student's chance of passing the test. Dependencies of a certain kind among the many characteristics are discussed by Xiaofeng Ma *et al.* (2018), who presents an introduction to attribute relationships as well as guidelines for assessing IG. The identifiers have a value that might be characterized as their connection.

$$DelayInfoGain(D,B) = W_i * Infor_{gain*(D,B)} \tag{3.19}$$

Equation 3.19 i describes the position of the major features, and W_i indicates the weight value of the variable i. DT is suited using the grid-based search technique. It is a system of improving the model performance by crossing a concerned group of identifiers.

3.8 Random Forest (RF)

RF is a common classifier, and it is a combination of various DT's performance. It is also used to forecast the value of the identifier. Every tree in the forest provides the votes and the classification of the class. RF selects the categorization having the best vote value. This regression forecasting has the following form.

Once the RF accepts the input vector value (x), which contains the values of various suggestive attributes analyzed for the concerned training region, it develops a K number of regression-type trees and produces the mean value of the outcomes. Afterward, T(x) trees are developed. Then, the split identifier was selected using GR, but it does not denote the major variations between the other choices using GI or IG.

3.8.1 Logistic Regression (LR)

The usage of the LR classifiers to guess the discrete-type values depends on a group of independent-type identifiers. It is called logit regression because it forecasts the chances of the existence of any type of event using the right information into a logit-type function. It provides linear-type class border values. Because of this reason, LR uses an S-type curve diagram in its place of the straight line. It is usually used to split data into various categories.

In student career prediction, the final field represents the objective of the model. The objective of the career prediction model is to find out the student's graduation level. The SAG solver is offered as it decreases the negative value of the Log-Likelihood method and provides the regularization feature. It also trusts the knowledge of gradient function. Here, the learning level is fixed in the step size 0.2. At last, the final constraints are set at the higher value of epochs and the Epsilon.

3.8.2 C4.5 Algorithm

According to Lokesh S. Katore *et al.* (2015), C4.5 is a supervised-type learning method, and the latest version is Id3. This concept is imitated by the scientist Ross Quinlan. The Id3 concept uses the value of IE. Here, the training information is divided by using IG. High-frequency rate attributes are measured for dividing the information depending on the data that existed in the given dataset. The following steps are used to predict the student's career performance using the C4.5 approach.

a) The base cases are checked.
b) Identify the normalized IG for the suitable split.
c) Arrange the data based on IG.
d) Make the DT.
e) Forecast the student's career.

Merits of C4.5

a) It manages discrete- and continuous-type values.
b) Missing attributes are also managed, and it will not reflect when computing the IG.
c) After formation, the pruning process is also executed. In C4.5, the designed score formula is illustrated below:

$$\sum z = \sum X_1C_1 + \sum X_2C_2 + \sum X_3C_3 \ldots\ldots\ldots\ldots + \sum X_nC_n \tag{3.20}$$

The output value of the career may be described by using Equation 3.20 z. The identifier of the person's personality is denoted by Xn, while the weight value of the career is denoted by Cn. This chapter analyzes and compares all of the currently available algorithms that fall under the umbrella of data mining and machine learning. Both the positive and negative aspects of each method are discussed in detail.

3.9 Conclusions

Applications based on graphs are present everywhere nowadays from knowledge graphs to social networks. Convolutional neural networks, often known as CNNs, have shown to be effective in a variety of fields, and their functionality may be extended to graph convolutional networks (GCNs). The graph Fourier transform is what is used to describe convolution when it comes to graphs. When it comes to defining the graph Fourier transform, the projection on the eigenvalues of the Laplacian is the way to go. These are what the graph refers to as its "vibration modes." A GCN, similar to typical CNNs, is made up of numerous layers that are convolutional and pooling to extract features followed by layers that are fully linked to complete the network. We make use of a polynomial parametrization to guarantee that the convolutional filters have compact support. Chebyshev polynomials make it possible to decrease the amount of complicated computing work.

References

1. Hechtlinger, Y., Chakravarti, P., Qin, J., *A generalization of convolutional neural networks to graph-structured data*, Cornell University Library, USA, 2017, *arXiv preprint arXiv:1704.08165.*
2. Zhao, W., Xu, C., Cui, Z., Zhang, T., Jiang, J., Zhang, Z., Yang, J., *When work matters: Transforming classical network structures to graph CNN*, Cornell University Library, USA, 2018, *arXiv preprint arXiv:1807.02653.*
3. De Plaen, P.F., Saerens, M., Lebichot, B., *Study of convolutional models for semi-supervised classification on graph-structured data*, Doctoral dissertation, MA thesis, Université catholique de Louvain, Belgium, 2021.
4. Zhang, Q., Chang, J., Meng, G., Xu, S., Xiang, S., Pan, C., Learning graph structure *via* graph convolutional networks. *Pattern Recognit.*, 95, 308–318, 2019.
5. Ma, Y., Wang, S., Aggarwal, C.C., Tang, J., Graph convolutional networks with eigenpooling, in: *Proceedings of the 25th ACM SIGKDD international conference on knowledge discovery & data mining*, pp. 723–731, July 2019.
6. Anirudh, R. and Thiagarajan, J.J., Bootstrapping graph convolutional neural networks for autism spectrum disorder classification, in: *ICASSP 2019-2019 IEEE International Conference on Acoustics, Speech and Signal Processing (ICASSP)*, IEEE, pp. 3197–3201, May 2019.
7. Yun, S., Kim, S., Lee, J., Kang, J., Kim, H.J., Neo-GNNs: Neighborhood overlap-aware graph neural networks for link prediction. *Adv. Neural Inf. Process. Syst.*, 34, 13683–13694, 2021.
8. Vashishth, S., Sanyal, S., Nitin, V., Talukdar, P., *Composition-based multi-relational graph convolutional networks*, Cornell University Library, USA, 2019, *arXiv preprint arXiv:1911.03082.*
9. Zhou, Z. and Li, X., *Graph convolution: A high-order and adaptive approach*, Cornell University Library, USA, 2017, *arXiv preprint arXiv:1706.09916.*
10. Zhao, T., Dong, N.T., Hanjalic, A., Khosla, M., *Multi-label Node Classification On Graph-Structured Data*, Cornell University Library, USA, 2023, *arXiv preprint arXiv:2304.10398.*
11. Jiang, X., Zhu, R., Li, S., Ji, P., Co-embedding of nodes and edges with graph neural networks. *IEEE Trans. Pattern Anal. Mach. Intell.*, 42, 388–401, 2020.
12. Peng, Z., Huang, W., Luo, M., Zheng, Q., Rong, Y., Xu, T., Huang, J., Graph representation learning *via* graphical mutual information maximization, in: *Proceedings of The Web Conference 2020*, pp. 259–270, April 2020.
13. Ge, K., Zhao, J.Q., Zhao, Y.Y., Gr-GNN: Gated recursion-based graph neural network algorithm. *Mathematics*, 10, 7, 1171, 2022.

14. Dernbach, S., Mohseni-Kabir, A., Pal, S., Towsley, D., Quantum walk neural networks for graph-structured data, in: *Complex Networks and Their Applications VII: Volume 2 Proceedings The 7th International Conference on Complex Networks and Their Applications Complex Networks 2018*, vol. 7, Springer International Publishing, pp. 182–193, 2019.
15. Liao, W., Bak-Jensen, B., Pillai, J.R., Wang, Y., Wang, Y., A review of graph neural networks and their applications in power systems. *J. Mod. Power Syst. Clean. Energy*, 10, 2, 345–360, 2021.
16. Zhou, J., Cui, G., Hu, S., Zhang, Z., Yang, C., Liu, Z., Sun, M., Graph neural networks: A review of methods and applications. *AI Open*, 1, 57–81, 2020.
17. Baron, M., *Topology and Prediction Focused Research on Graph Convolutional Neural Networks*, Cornell University Library, USA, 2018, *arXiv preprint arXiv:1808.07769.*
18. Wu, M. and Rostami, M., *Unsupervised Domain Adaptation for Graph-Structured Data Using Class-Conditional Distribution Alignment*, Cornell University Library, USA, 2023, *arXiv preprint arXiv:2301.12361.*
19. Kim, R., So, C.H., Jeong, M., Lee, S., Kim, J., Kang, J., *Hats: A hierarchical graph attention network for stock movement prediction*, Cornell University Library, USA, 2019, *arXiv preprint arXiv:1908.07999.*
20. Huang, J., Li, Z., Li, N., Liu, S., Li, G., Attpool: Towards hierarchical feature representation in graph convolutional networks *via* attention mechanism, in: *Proceedings of the IEEE/CVF international conference on computer vision*, pp. 6480–6489, 2019.
21. Li, C.Y., Liang, X., Hu, Z., Xing, E.P., Knowledge-driven encode, retrieve, paraphrase for medical image report generation, in: *Proceedings of the AAAI Conference on Artificial Intelligence*, vol. 33, pp. 6666–6673, July 2019.
22. Li, R., Wang, S., Zhu, F., Huang, J., Adaptive graph convolutional neural networks, in: *Proceedings of the AAAI conference on artificial intelligence*, April 2018.
23. Liu, G., Huang, X., Yi, X., Adversarial Label Poisoning Attack on Graph Neural Networks *via* Label Propagation, in: *European Conference on Computer Vision*, Springer Nature Switzerland, Cham, pp. 227–243, October 2022.
24. Pradhyumna, P. and Shreya, G.P., Graph neural network (GNN) in image and video understanding using deep learning for computer vision applications, in: *2021 Second International Conference on Electronics and Sustainable Communication Systems (ICESC)*, IEEE, pp. 1183–1189, August 2021.

25. Andreoletti, D., Troia, S., Musumeci, F., Giordano, S., Maier, G., Tornatore, M., Network traffic prediction based on diffusion convolutional recurrent neural networks, in: *IEEE INFOCOM 2019-IEEE Conference on Computer Communications Workshops (INFOCOM WKSHPS)*, IEEE, pp. 246–251, April 2019.

4

Research Trends and Challenges of GCNN Over CNN and Digital Image Processing Techniques

Rithish Kanna S.[1*], Suganthi P.[1] and Kavitha P.[2]

[1]*Department of Computer Science and Business Systems, Thiagarajar College of Engineering, Madurai, India*
[2]*Department of Information Technology, Velammal College of Engineering and Technology, Madurai, India*

Abstract
By fusing a content image with a style image using an efficient method known as Neural Style Transfer (NST), a new image is generated that preserves the original content while adopting the artistic style of the reference image. Artists and designers prefer NST because it offers a faster alternative to manual stylization. The segment-specific neural style transfer techniques that can be applied are the main topic of this work. Hybrid neural style transfer allows previously generated styles to be reused and selectively applied to different regions of an image. There are different areas of the image, each of which can employ a different aesthetic. This approach enhances the artistic quality of the resulting image. The content image defines the structure and objects in the scene, while the style image provides the visual appearance or artistic characteristics to be applied. This technique lies at the intersection of computer vision, convolutional neural networks, and image processing. Globally, this technique has been employed by designers and artists to create new works of art based on existing trends.

***Keywords*:** Artificial intelligence, convolutional neural network, deep learning, hybrid neural style transfer, machine learning, neural style transfer, style transfer, content image

**Corresponding author*: rithish1122004@gmail.com

Malini Alagarsamy, Rajesh Kumar Dhanaraj, J. Felicia Lilian, Vandana Sharma and Gheorghita Ghinea (eds.) Graph Convolutional Neural Networks for Computer Vision, (73–94)

4.1 Introduction

Artificial intelligence (AI) is a computer- or robot-controlled domain that is capable of performing the same task as done by humans. AI mainly falls into two main categories, namely, machine learning and deep learning. Complex problems are solved with the techniques of machine learning. Machine learning (ML) provides outputs based on structured data algorithms. Deep learning, which relies on complex neural networks, mimics human intelligence in recognizing patterns and learning from datasets. The database can be either structural or nonstructural. Image processing is usually done using deep learning. Intelligent Document Processing (IDP) uses deep learning networks known as CNNs (Convolutional Neural Networks) to learn patterns that occur naturally in photographs. These are used mainly to convert unstructured data. The artistic style of one picture is applied to the original content of some other picture, which is the basic technique used in NST. This process can be used to achieve various artistic effects. This idea not only allows more artistic control over the final image but also provides an approach to generate a set of different stylized images from a single source image. In addition, neural style transfer helps to give depth and realism to grayscale images. Evaluation is done through a series of experiments and demonstrates its effectiveness in providing the best-quality stylized images using multiple segments and coloring. Also compared our implementation of neural style transfer with other current models to demonstrate its strengths. Style Transfer (ST), or otherwise redrawing artwork, is a tricky subject. Nevertheless, this paper focuses on a specialized, structured deep CNN architecture that is effectively used for style transfer. A Convolutional Neural Network (CNN) is a deep learning framework that learns patterns and features directly from data.

CNNs are especially used in changing patterns to match orders, objects, and classes in images. It is also effective in classifying signal data, video data, audio data, and time series data. Two images are used as network inputs. They are content images and style images, and the output image thus obtained is generated image. The algorithms and datasets used in ML research are quickly picked up by artists to innovate playful, provocative, and boundary-pushing art. This allows you to enhance images shot in sub-optimal conditions with images that have the perfect combination of color correction, lighting, and contrast. Another contribution is the optimization of transferred images by improving image aesthetics with the help of neural image assessment. By using the neural style transfer technique, we can tackle the impossible that we cannot even imagine. Similarly, you can also combine two or more images. It also helps you see things beyond your imagination.

Overall, the goal is to demonstrate the potential of NST for applications and obtain a useful approach for generating a variety of stylized images.

4.1.1 Frequently Used Artistic Styles

Abstract, figurative, geometric, minimalist, nature, pop, portrait, still life, surrealist, typography, and urban are various styles used by artists. There are various mediums, which the artists can use in making various artworks. Some commonly used mediums include oil paints, colored pencils, charcoal pencils, graphite pencils, chalk, crayons, watercolors, dry media, and spray paints where in some cases, heat plays a role in the application or drying process. Art work done by spray paint provides a realistic look. Graphite pencils come in various shades that produce different levels of darkness, with HB pencils being the most commonly used type today.

4.2 Introduction to Convolutional Neural Network

The NST uses the CNN to provide a turnout image with applied styles. Artificial neural networks are used to process words, images, and audio. There are different types of neural networks. One such neural network is Recurrent Neural Networks, which is used to predict the sequence of words, and there are various projects done using RCNN. One such work is an automatic assessment of road conditions from aerial imagery using VGG architecture in the faster-RCNN framework [6].

CNN is a notable type of ANN (Artificial Neural Network) in which the connectivity patterns between the neurons are due to the visual cortex. The visual cortex is a small region of cells that are used in visual fields. Some of these neuron cells respond at the edges, and some do not. This is the basic concept behind the neural networks.

Convolutional Neural Networks (CNNs), also known as ConvNets, are a type of neural network where parameters are shared using convolutional filters. This weight sharing helps reduce the number of parameters and allows the network to detect patterns efficiently. The CNN has multiple layers. One of the advantages of CNN is that there is no need to do much processing on the images. This technology does not require much trial and error and does not require many parameters. The filters are typically created using heuristics. The main purpose of the CNN is to convert data into forms that are easier to process with minimal loss of features.

The main difference between CNNs and regular neural networks is that CNNs use convolutions to provide the exact output. They take two input functions and return output function. The convolutions are used instead of matrix multiplication in at least one of its works. It is possible to tune the filters using CNNs. The CNNs are based on neuroscience. As discussed earlier, they contain nodes that calculate the weighted sum of inputs and provide an activated map. When an image is uploaded, it takes up pixel values and picks some of the visual features. These will point to the pixel values like colors and give the output activation map. The starting point of CNN is nothing but locating the borders of the image. The layer will then receive this minor image definition and begin detecting things like corners and color groups. As the subsequent layer receives the image definition, it is repeated in a cycle until the entire predictions are made. The layer becomes more defined; it provides only the relevant features, and they move on to the succeeding layer. This process keeps up until the final coating is achieved. An activation map for each layer is produced. The last layer in the CNN classification produces the final predicted values based on the activation map. CNNs are widely used in applications like autonomous vehicles, where they help recognize and classify objects such as cars, buses, and pedestrians.

There are different types of CNN available, and a short description of each of them is shown below.

4.2.1 Types of CNN

4.2.1.1 1D CNN

One-dimensional CNNs modify the standard CNN kernel to operate over a single spatial dimension. They are commonly used for analyzing time-series data.

4.2.1.2 2D CNN

The two-dimensional CNN usually moves in a two-dimensional network, and the common applications where 2D CNN is used are image processing and image labeling. These are widely used in image recognition with little preprocessing, recognizing various styles of handwriting, computer vision software, reading check digits in banking, and reading zip codes on an envelope at the postal service.

4.2.1.3 3D CNN

The three-dimensional CNN moves in three dimensional space, which has a kernel. These are highly advantageous and are widely used in the medical fields such as MRIs and CT scans.

The work of CNN is thus discussed. There is a mathematical way of solving the CNN problem, but this paper does not deal with it. This paper focuses only on the introduction of CNN.

4.2.2 Application of CNN

The following examples illustrate common applications of CNNs.

4.2.2.1 Face Detection

CNNs are used for the detection of faces in photographic or portrait pictures. By receiving a picture as data from the user, the network provides an output as a set of values that indicate the attribute's facial trait at various bits in the picture. According to numerous research publications, CNNs have demonstrated higher accuracy than earlier algorithms, correctly recognizing faces 97% of the time. CNN can be used to lessen the degree of facial bent. CNNs are quite accurate at identifying facial characteristics like the eyes, nose, and mouth, which helps to minimize distortions caused by factors like angles or shadows.

4.2.2.2 Object Detection

CNN has been used to classify items based on criteria of patterns and shapes present in a picture to recognize objects across images. CNNs have been used to develop models capable of recognizing a wide range of objects, including everyday items, popular foods, animals, birds, currency (like dollar or euro bills), and even unusual items such as firearms. Techniques like semantic or instance segmentation are used for object detection. For use in drones or self-driving automobiles, CNNs have been used to locate and find things in photos as well as to create various perspectives of those objects [18].

4.2.2.3 X-Ray Image Analysis

To find tumors or other anomalies in X-ray pictures, CNNs have been utilized in medical imaging. Based on previously saved images, they are trained.

Then, the process of similar images by CNN networks takes place. CNN models can analyze images of the human body and identify specific areas, such as locating a tumor near the knee or detecting other abnormalities. They are also widely used to examine medical images like X-rays to identify anomalies. CNNs have proven effective in pinpointing the location of tumors and other medical conditions with high accuracy [19].

4.2.2.4 Other Applications

The concept of CNN is also used in various applications, which are listed below.

- Facial emotional recognition
- Self-driving or autonomous cars
- Automatic translation
- Prediction of the next word in a sentence
- Recognition of handwritten character
- Cancer detection
- Captioning an image
- Answering visual questions
- Authentication of biometrics
- Document classification
- Segmentation of 3D medical image

4.3 Neural Style Transfer—Artistic View

A group of software algorithms known as Neural Style Transfer (NST) alters videos or digitalized images to take on the visualization style of another image. Deep neural networks are a feature of NST algorithms, which they use to modify images. NST is frequently used to create false art from photographs, which is the main content. For example, it can transform user-supplied photos to resemble well-known paintings. NST methods are employed for this purpose by several well-known smart phone apps, including Deep Art and Prisma. Artists and designers have used this technique to create new works of art based on existing styles all around the world.

There are typically two kinds of style transfer. One type of style transfer is referred to as "photorealistic," in which the text and style images are both "genuine." A photorealistic transfer aims to enhance or add anything to the original content image. The next type of style transfer is cited as an

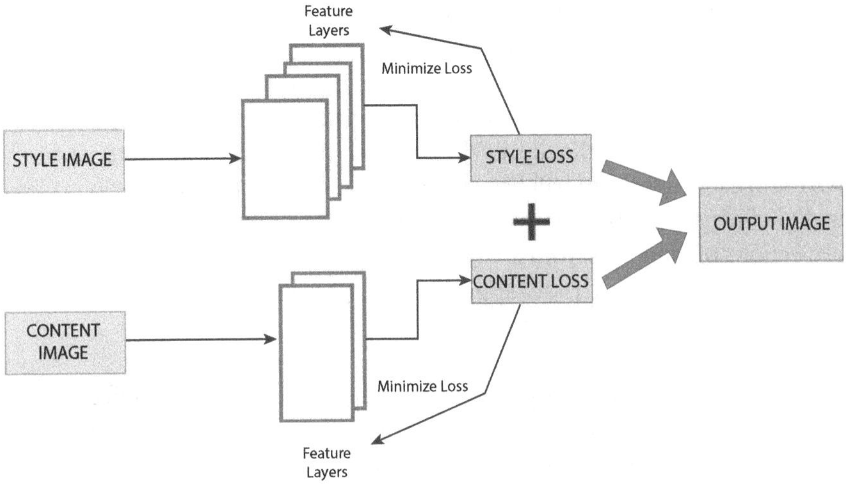

Figure 4.1 Block diagram for NST.

"artistic" style transfer, in which the artistic design of a particular picture or painting is applied to the content image. "Feature maps" are used to represent content information. These are matrices that include the image's edges and any other higher-level features that may be present. Thanks to these developments in modern technology, nearly everyone can now enjoy the joy of creating and disseminating great, beautiful artwork.

This is where the ability of style transfer to change lies. The coexistence of contemporary works with timeless classics is made possible by artists' capacity to pass on their style to others. The importance of neural style transfer in the commercial art industry is evident, and it has inspired millions of people worldwide to pursue their creative endeavors. The NST is a technique content is added to the style image to generate a final output image as shown in Figure 4.1. This neural style transfer is an optimistic approach and is used in various fields.

4.4 Various Existing Works of NST

The choice of the appropriate style for an image is the artist's current dilemma. This is typical of the majority of artists, and choosing the media to utilize can often be challenging. Painting, sketching, oil pasting, and other different mediums are examples. There have been numerous recent advancements in the situation, such as spray painting, which has just been completed and needs heat to dry. Different goods have developed over time to utilize various mediums. The problem statement mainly focuses on how the artist feels confused in choosing different styles mentioned in the introduction. The improvement of various style transfer methods requires considering factors beyond quantitative evaluation, as outlined in [9]. There are various algorithms to perform NST and each algorithm will have various efficiency factors [6]. Various ideas include the usage of brushstrokes rather than pixels to style the image [7] and projections of emotions from artworks to maps using NST [13].

There exist multiple tools/applications to use neural style transfer algorithms to create an artistic image.

- Prisma is an editing photo app that provides amazing effects on photos that are attractive to the users and performs various transformations of photographic pictures to paintings. Prisma makes use of various artificial neural networks that allow users to make their pictures appear as they were portrayed by "Picasso" or "Salvador Dali" himself.
- The mobile app "Wizart" uses amazing artworks using deep neural style algorithm technologies. The style utilizes neural networks and AI to take an image or collection of images or even videos that you provide and apply an existing graphical style to your media.
- In addition, some of the other applications, like Pixify and Artisto, use NST.

The paper [15] provides a semantically accurate generated image for any content-style image input pair and discusses the unidirectional-GAN model.

4.4.1 Literature Survey

The papers that follow the neural style transfer concepts are listed below.

- Applying the approach of NST to work on videos: The video is separated frame by frame, and then, the NST is applied to each frame of the video. The feed-forward convolution neural network is used where the output is both stylized and temporarily consistent. The loss of temporal coherence during the training stage is calculated using a two-frame synergic training mechanism [1].
- The mathematical approach of the gram matrix does not provide accurate results. The mean discrepancy at maximum

uses the kernel's second-order polynomial to provide accurate results [2].

- Starting with a random-noise input signal, a style transfer problem is solved iteratively by optimizing the noise to match sieve achievement from an interesting previously trained neural network. Examining two distinct activities that result in compression and expansion of bandwidth transfer from voices, like singing, to musical instruments are examples [3].
- The instabilities of a few algorithms are highlighted, and it is demonstrated that these flaws can be exploited to increase the variety and appealing weirdness of the pictures produced by the several NST techniques now in use. The computer was shown as a computational catalyst providing human painters with a variety of images that served as a source of inspiration [4].
- Interactive image editing using MetalAPI and CoreML [16, 17].
- Composition, projection technique, and object wrapping and bending are all components of the style. Introducing a neural architecture that facilitates geometric style transfer using semantic content [5].
- Rectifying the two types of losses when a style is applied, namely, fixpoint triplet style loss and disentanglement loss [8].
- Improving the content loss using a deep encoder network for real-time stylized images and videos [10].
- Use of residual Unet for grayscale images and the usage of NST in anime images [11].
- Automating motion style transfer, which will save the work of the animators [12].
- Usage of CNN for Chinese paintings and proposal of the algorithm of Chinese painting style transfer [14].

4.5 Hybrid Neural Style Transfer

The NST technique, as previously mentioned, generates a final output picture by combining the image's style with the main content image. The content image is first divided into two halves. The split halves are of the same size. The outputs of the first and second parts, along with their respective styles and the images, are now fed as input to the merging algorithm.

Style 1 is applied to the first half of the image during the first iteration. Style 2 is applied to the second part of the image during the second iteration. The first and second halves of the image are combined during the merging process, so there will be no problems because the image split earlier will be of the same width and length. Style 1 on the left and style 2 on the right of an image will be precisely merged into the final output.

Artists who are unsure of their preferred style or medium can make effective use of this procedure. The artist can simply upload a content image and two styled images, where the styles need to be applied. The process takes some time and produces the final image by merging image 1 with style and image 2 with style. For more information about styles and mediums, see the Introduction section, which was discussed earlier in section 4.1. There are two halves in the intermediate results. Style 1 uses the image's left half, and style 2 uses the right half. As previously stated, the current loss value is relevant here. However, although it takes more time, the precise merging of style and image is achieved, which is significantly more effective than the usual results. Finally, the HNST assists the artist in determining whether the two distinct styles complement the image simultaneously. This concept can be extended by splitting the image into 4, 8, 16, or more equal parts. Different styles can then be applied to each part simultaneously. However, the effectiveness of this approach depends on the nature of the artwork. some styles may not blend well when applied in separate sections. From my perspective, there are two key scenarios that explain this limitation.

4.5.1 Scenarios Considered

- **First Scenario:** If the portrait is centrally fixed, the left eyes, left part of the nose, left eyebrows, left lip, etc., will be applied with one style, and the parts on the right will have another style making it easy to compare. However, if the portrait is divided into four halves, it looks like four quadrants of the graph with the first part on the upper right, second part on the upper left, and the third and fourth parts on the lower left and lower right, respectively, will all have distinct styles making it difficult to compare. A similar problem exists when the portrait is divided into 16, 32 parts, etc.

- **Second Scenario:** Consider a landscape image. If the image is centrally fixed, the left mountain, left part of the river, left part of the sky, left part of the land, etc., will have one style, and the parts on the right will have other styles making it easy to compare. In a similar vein, this issue exists when the image is divided into 16 and 32 parts. Because of this, we restricted the splitting to two halves. Additionally, as the number increases, the merging algorithm becomes more complicated. The block diagram of HNST is shown in Figure 4.2, and its flow is illustrated in Figure 4.3.

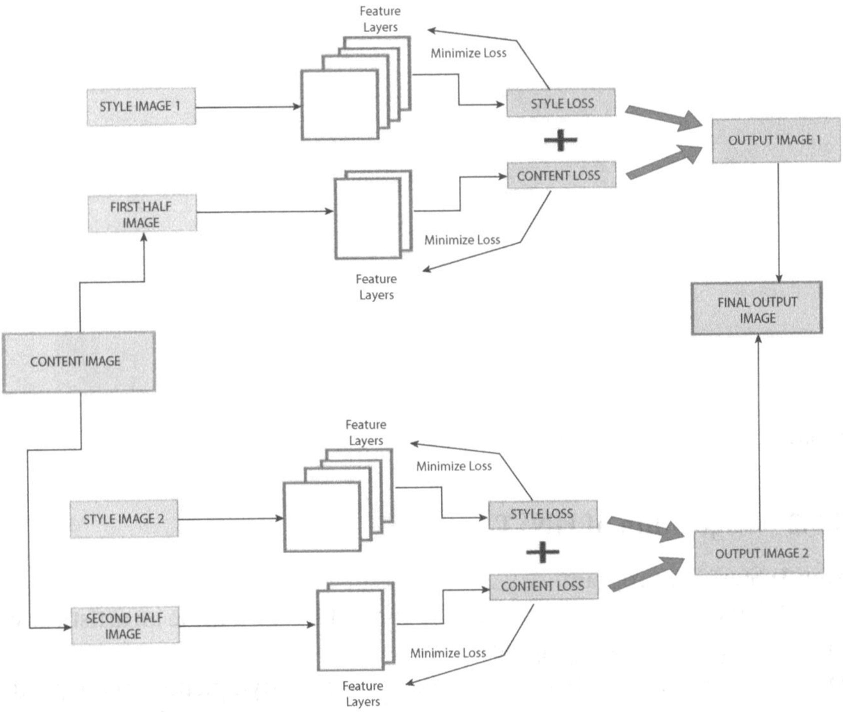

Figure 4.2 HNST's block diagram.

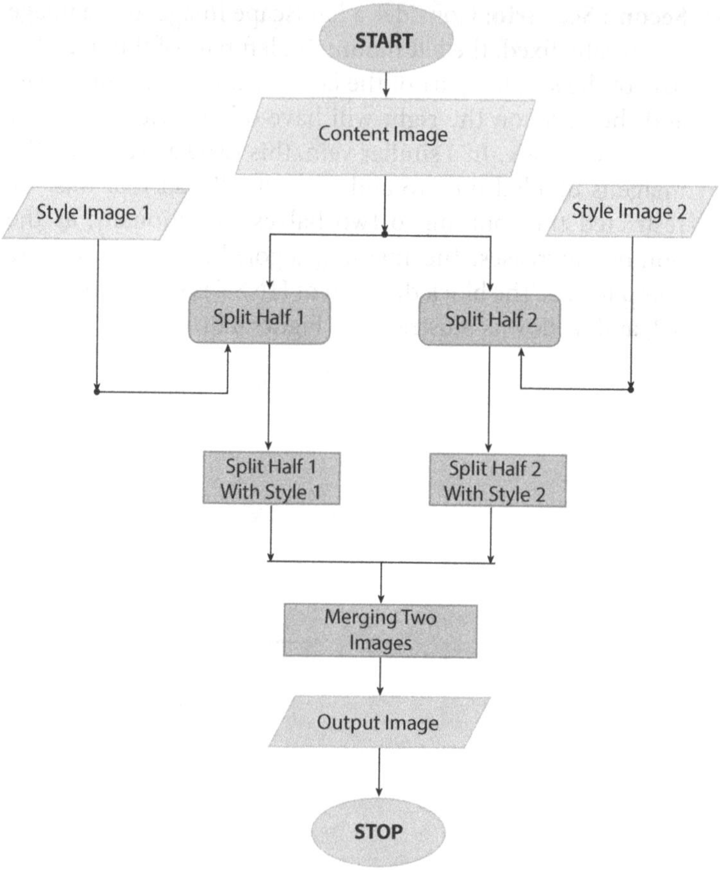

Figure 4.3 Flowchart of HNST.

4.5.2 Algorithm of HNST

Step 1: Start the program.
Step 2: Upload the content image. The content image may be an artistically drawn image or a photographic image.
Step 3: Uploading of the first style image and second style picture. The applied stylized image can either be the style itself or an image where a blend of styles has been applied.
Step 4: Separation of the content image into two halves, namely, split half 1 and split half 2 (content image = split half 1 + split half 2).

Step 5: Applying NST where the content image is split half 1 and the image's style is style 1, which produces the output image 1.
Step 6: Applying NST where the content image is split in half 2 and the image's style is style 2, which produces the output image 2.
Step 7: Merging of produced image 1 with the produced picture 2 to generate the final output image.
Step 8: The Final output image can be downloaded.
Step 9: End of the program.

4.6 Implementation of HNST

The hybrid neural style transfer is implemented using Python programming. There are various deep-learning libraries available in Python that deal with the images. So this process of HNST can be implemented using the Python libraries that deal with image processing and deep learning algorithms. The usage of some of the important Python Libraries on the images is shown below.

4.6.1 OpenCV

An OpenCV is a collection of Python bindings. Python was created to address various issues regarding computer vision. A window containing an image is displayed using the cv2.imshow() technique. The image size is automatically adjusted for the window. OpenCV makes it possible to carry out various tasks of image processing and computer vision.

4.6.2 Pillow

One of the best key modules in the Python image processing modules present is PIL. A lot of features are available in the pillow module and is suited to all famous operating systems. The PIL modules support a wide variety of image formats, including "jpeg," "png," "gif," and "tiff." Using the PIL module, you will be able to perform anything with digitalized photos uploaded.

4.6.3 Imageio

ImageIo is a simple interface, especially for writing and reading a variety of image data, which includes animated images, scientific formats, and volumetric images. It is platform-neutral Imageio. imwrite() must be used to create an image file. We must use the image.get_reader() method to read a file.

4.6.4 TensorFlow

TensorFlow is a Python library that is used to perform quick numerical computations, and this was developed and released by Google. It is a base library that is basically used to create deep learning models indirectly or even directly using libraries that are wrapped and created on TensorFlow to make each step involved simpler. TensorFlow is an open-root, Python-accordant toolkit for algebraic computation that processes and simplifies the formation of networks of neural and ML models.

4.7 Results and Inference

4.7.1 For Sketch Images

Figure 4.4 illustrates the selected content image along with the two style images, which are used as inputs for the Hybrid Neural Style Transfer (HNST) process.

Figure 4.4 Selection of images.

Figure 4.5 demonstrates how the image is split into left and right halves. In Figure 4.6, Style 1 is applied to the left half, while Figure 4.7 shows the application of Style 2 to the right half. Finally, Figure 4.8 presents the combined image after both styles have been applied to their respective halves.

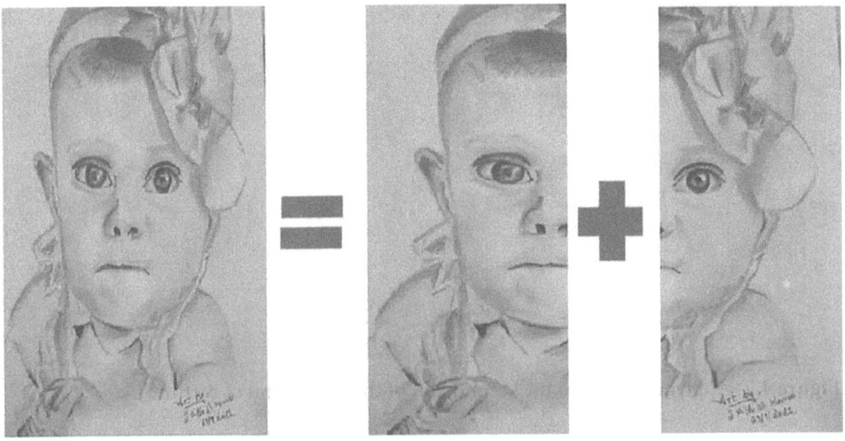

Figure 4.5 Separation of content image.

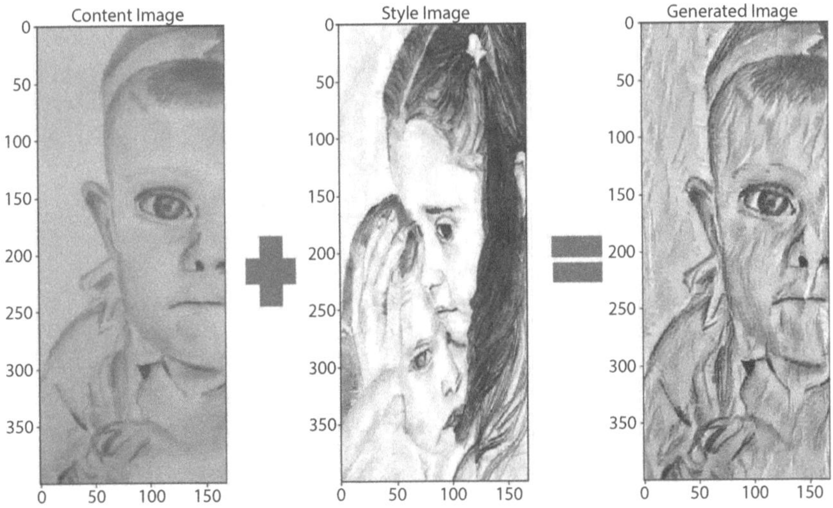

Figure 4.6 Applying style 1 to the first half of the main content picture.

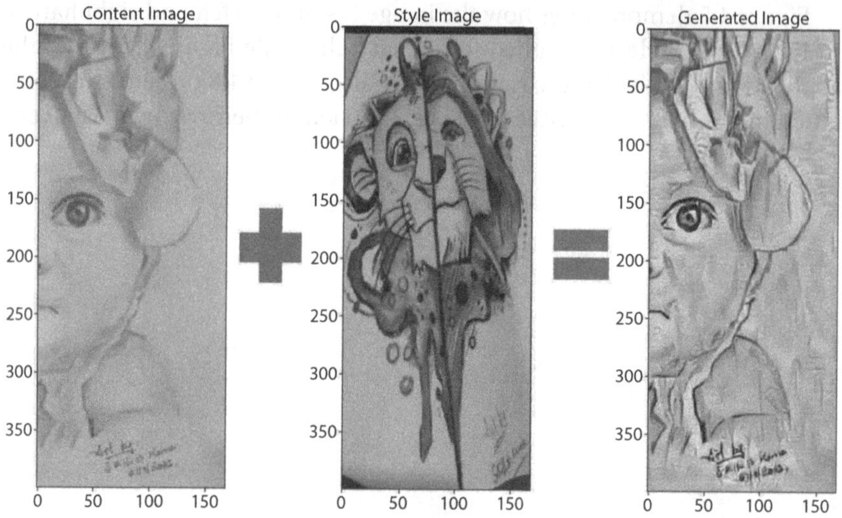

Figure 4.7 Applying style 2 to the second half of the main content picture.

Figure 4.8 Final output image figure.

The above is the output produced based on the drawings. The drawings include the medium pencil shading and colored pencils. Here, the content image is the pencil shading image. Style 1 is a pencil shading image, whereas style 2 is a color pencil image. In the final output image, you can note that the second half (right part of the image) has certain colors applied to it.

4.7.2 For Photographic Image

Figure 4.9 illustrates the selection of styles for the photographic content image. In Figure 4.10, the image is split into left and right halves. Figure 4.11 shows the application of Style 1 to the left half of the photographic image

Figure 4.9 Selection of images.

Figure 4.10 Separation of content image.

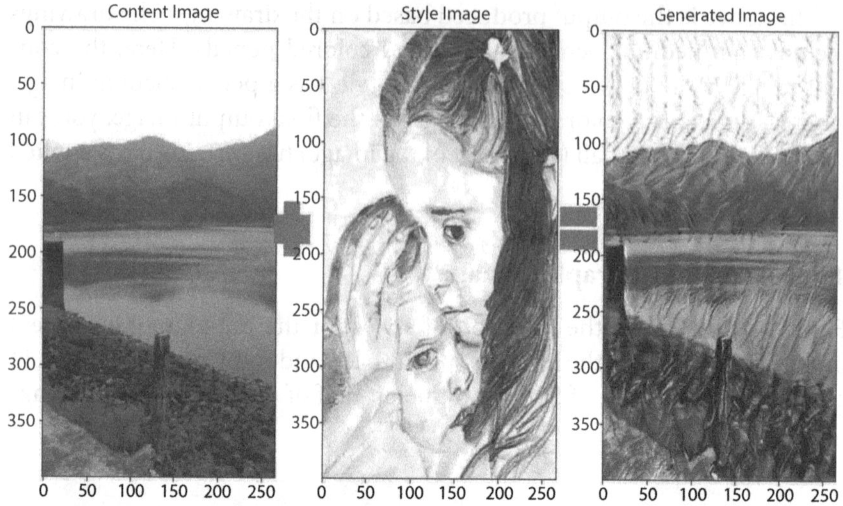

Figure 4.11 Applying style 1 to the first half of the content image.

and Style 2 to the right half, as demonstrated in Figure 4.12. Finally, Figure 4.13 presents the final output image after the styles have been applied.

By applying HNST to the photographic image's main content, it is the nature picture that is photographed, and the styles of drawings are applied

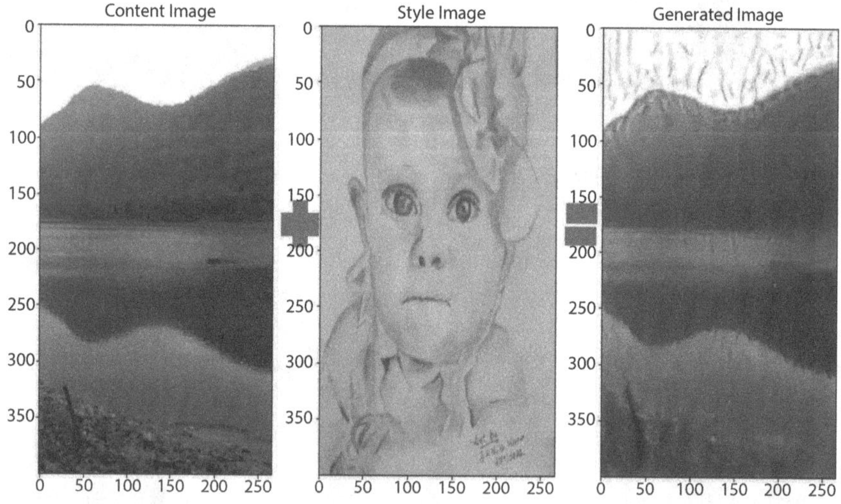

Figure 4.12 Applying style 2 to the second half of the content image.

Figure 4.13 Final output image.

to check which styles suit better. In the above example, the left half produces a better highlighted view than the right half. According to my view, the left-half style can be applied so that the final output image looks better. This helps us make comparisons and choose the best style among these two styles.

4.8 Further Ideas of HNST

4.8.1 Introduction to HNST 2

In HNST 1, we used three types of images, the content image and two style images. The image was then split into two halves, NST was applied to the first and second halves separately, and the results were then merged to produce the final output image. Here, the same HNST approach is used to derive HNST 2 with a slight modification to the existing one. As discussed in HNST, the content image is first separated into two parts, but not into two halves. Instead, the background is separated as the first part, and the main image is separated as the second part as shown in Figure 4.14. The NST is applied to the two different separated halves, and finally, the merging up of images takes place to produce the final output image.

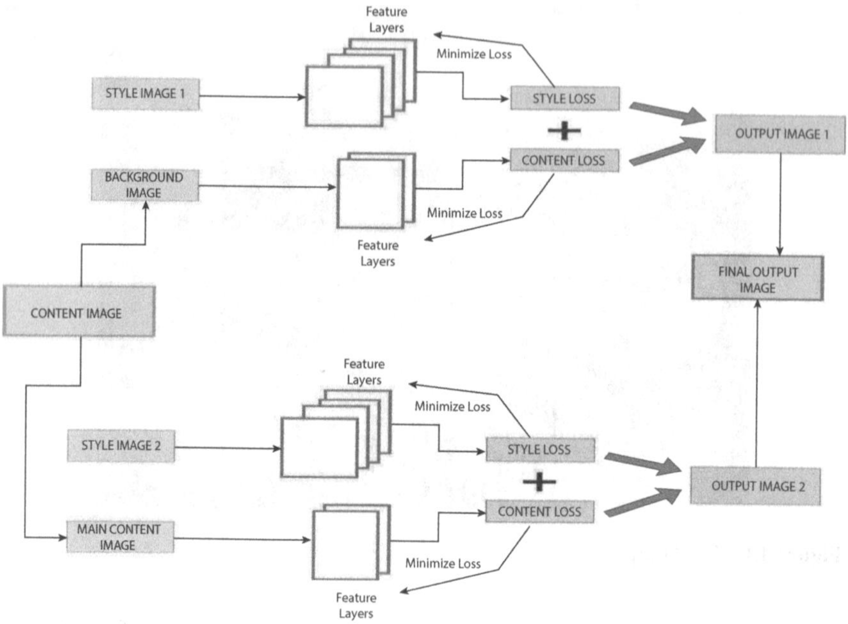

Figure 4.14 Block diagram for HNST 2.

4.9 Conclusion

Artificial intelligence's evolution is gaining popularity and is valuable in both the present and the future. This paper discussed the CNN's introduction and the numerous applications in which it is mostly used. In CNN's primary classification, 3D CNN is progressively offering a bunch of practical uses. The focus of this chapter is to present an artistic perspective of using NST, which enables people to provide a more detailed piece of art with appropriate medium and style selection. The existing approaches covered in the NST techniques are explored. Finally, a thorough explanation of the HNST introduction and its working was provided.

References

1. Huang, H., Wang, H., Luo, W., Ma, L., Jiang, W., Real-Time Neural Style Transfer for Videos. *Proceedings of the IEEE Conference on Computer Vision and Pattern Recognition (CVPR)*, 2017.
2. Li, Y., Wang, N., Liu, J., Hou, X., Demystifying Neural Style Transfer. *IJCAI*, 2017.

3. Verma, P. and Smith, J.O., 'Neural Style Transfer for Audio Spectograms. *31st Conference on Neural Information Processing Systems (NIPS)*, Long Beach, CA, USA at the workshop for Machine Learning for Creativity and Design, 2017.
4. Kerdreux, T., Thiry, L., Kerdreux, E., *Interactive Neural Style Transfer with Artists*, 2020.
5. Liu, X.-C., Li, X.-Y., Cheng, M.-M., Hall, P., Geometric Style Transfer, 2020, *arXiv preprint arXiv:2007.05471.*
6. Malini, A., Priyadharshini, P., Sabeena, S., An automatic assessment of road condition from aerial imagery using modified VGG architecture in faster-RCNN framework. *J. Int. Fuzzy Syst.*, 40, 6, 11411–11422, 2021.
7. Kotovenko, D., Wright, M., Heimbrecht, A., Ommer, B., Rethinking Style Transfer:From Pixels to Parameterized Brushstrokes. *Proceedings of the IEEE/CVF Conference on Computer Vision and Pattern Recognition (CVPR)*, 2021.
8. Kotovenko, D., Sanakoyeu, A., Lang, S., Ommer, B., Content and Style Disentanglement for Artistic Style Transfer. *Proceedings of the IEEE/CVF International Conference on Computer Vision (ICCV)*, 2019.
9. Wright, M. and Ommer, B., *ArtFID: Quantitative Evaluation of Neural Style Transfer*, Springer, Cham publisher, 2022.
10. Sanakoyeu, A., Kotovenko, D., Lang, S., Ommer, B., A Style-Aware Content Loss for Real-time HD Style Transfer. *Proceedings of the European Conference on Computer Vision (ECCV)*, 2018.
11. Zhang, L., Ji, Y., Lin, X., Liu, C., *Style Transfer for Anime Sketches with Enhanced Residual U-net and Auxiliary Classifier GAN*, IEEE Publisher, 2017.
12. Holden, D., Habibie, I., Kusajima, I., Komura, T., Fast neural style transfer for motion data. *IEEE Comput. Graphics Appl.*, 37, 4, 42–49, 2017.
13. Bogucka, E.P. and Meng, L., Projecting emotions from artworks to maps using neural style transfer. *Proceedings of the International Cartographic Association*, vol. 2, 2019.
14. Sheng, J., Song, C., Wang, J., Han, Y., Convolutional neural network style transfer towards Chinese paintings. *IEEE Access*, 7, 163719–163728, 2019.
15. Deelaka, P. N. Neural Artistic Style Transfer with Conditional Adversaria, 2023. *arXiv preprint, arXiv:2302.03875.*
16. Paswaldt, S., Semmo, A., Klingbell, M., Dollner, J., *Demo: Victory-neural style transfer and editing with coreML*, Assosiation of Computing Machinery New York, NY, United States, 2017.
17. Basheer, S., Singh, K.U., Sharma, V., Bhatia, S., Pande, N., Kumar, A., A robust NIfTI image authentication framework to ensure reliable and safe diagnosis. *PeerJ Comput. Sci.*, 9, e1323, 2023.
18. M.N, R.D., Murali, S., Sharma, V., Performance Analysis of DGA-Driven Botnets using Artificial Neural networks. *2022 10th International Conference on*

Reliability, Infocom Technologies and Optimization (Trends and Future Directions) (ICRITO), pp. 1–6, 2022, doi: 10.1109/ICRITO56286.2022.9965044.

19. Sharma, V. and Midhunchakkaravarthy, D., SVM Based AutoEncoder for Detecting Dementia in Young Adults. *2023 6th International Conference on Contemporary Computing and Informatics (IC3I)*, Gautam Buddha Nagar, India, pp. 2516–2520, 2023, doi: 10.1109/IC3I59117.2023.10397637.

5

Classification of Graph Filtering Operations and Inductive Learning by Exploiting Multiple Graphs in GCNN

S. Kayalvizhi*, Harish Sekar and Prasanna Guptha M.P.

Department of Electronics and Communication Engineering, Faculty of Engineering and Technology, SRM Institute of Science and Technology, Kattankulathur, Tamil Nadu, India

Abstract

Graph filters are mathematical methods or techniques that are used to process and evaluate data represented as graphs. A mathematical structure made up of edges and the nodes that link them is called a graph. For many applications using graph data, including network research, social network analysis, and recommendation systems, graph filters are crucial tools. Graph filters are very similar to conventional filters. They are monotonic, shift constant, parametric operations of the input, possess a spectral understanding using spectral graph theory, and their spectral layout reduces to an adjusting function. Nonetheless, differing distinctions also emerge from the new graphical mediums; for instance, graph strainers correspond to permutations in support, have an inherently dispersed execution, and can adopt more generalized forms such as a variable of vertices or connections. Due to the large variance in network signals and the versatility of graphs to denote erratic structure in complex data domains, these are used in countless Signal Processing and Machine Learning applications.

Keywords: Graph filters, convolution filters, neural network, GCNN, inductive learning

**Corresponding author*: kayalvis@srmist.edu.in

Malini Alagarsamy, Rajesh Kumar Dhanaraj, J. Felicia Lilian, Vandana Sharma and Gheorghita Ghinea (eds.) Graph Convolutional Neural Networks for Computer Vision, (95–116) © 2026 Scrivener Publishing LLC

5.1 Introduction

Filters are architectures for processing the information; they preserve only the relevant input contents. Filtering involves preserving certain spectral components of the input and serves as a fundamental element in numerous areas of signal processing such as radar communications and multimedia [1]. Filtering is utilized to obtain pertinent trends from data in constructing artificial networks [2]. These techniques are applied to signals defined on Euclidean domains. Early graph filtering formalisms have their roots in network processing in the 1990s [3, 4]. Graph filters became popular in the 2000s first as graph kernels or regularizes in machine learning applications [5, 6] and later in graph-based image processing [7, 8]. Concerning signal processing, the initial one may be regarded as a graph wavelet [9, 10], while the second one is a generalized graph filter bank [11]. This tutorial paper helps to provide a unified mathematical foundation for numerous specific problems pursued in various scientific fields allowing more communication researchers to easily solve challenges related to ANNs and create differential filtering techniques. Encourages. The specialization of algebraic signal processing frameworks in the graph domain has made structured mathematical frameworks for graph signal processing (GSP) and graph filtering viable concurrently [12, 13]. Since graph neural networks (GNN) are the most widely used tools for graphical machine learning (GML), graph filters have become increasingly important as a key element in capturing relationships between data and its structure [14–16].

Our aim this chapter is providing the reader with an extensive information about different forms of graph filtering. The basics of the graph and its signals are discussed in Section 5.2, and Section 5.3 analyzes the various forms of graph filters and properties. Section 5.4 discusses the graph filter bank and wavelets, Section 5.5 deals with the graph neural networks, and Section 5.6 has the conclusion remarks.

5.2 Graph Basics

Graph filters are mathematical processes or methods for analyzing and interpreting data that are represented as graphs. A graph constitutes a form of mathematical construct made up of nodal points, vertices, and edges, or joins, which link the nodes together. Graph filters are essential tools for various applications that involve graph data, such as social system analysis, recommendations, and network science.

A. Graph Terms

There are two types of graphs: directed and undirected. The adjacency set for node i is $N_i = \{j \in V: (i, j) \in e\}$. In an undirected graph, the edges in e, $\omega(i, j) = \omega(j, i)$ have no direction. In a directional graph, an edge $(i, j) \in e$ has position originating at node i and terminating at node j. i's external neighbor is node j. $N_i^{in} = \{j \in V : (j, i) \in e\}$ indicates the adjacent set, while $N_i^{out} = \{j \in V : (i, j) \in e\}$ indicates the out-neighbor set of node i.

B. Graph Signals

Graph signals are processed using graph filters and GNNs. The vectorized graphical impulse is expressed as $\chi \in R_n$, where the graph's -th node is represented by component X_i. A graph containing a signal made up of nine nodes and nine components is displayed in Figure 5.1. Different network nodes correspond to different signal components. Signals have their own shift operators. A pair (S, x) can be used to represent a signal. Utilizing this known information during signal processing is the aim of graphical signal processing. The expected value plot is the proximity or similarity between signal components x.

When employing a shift operator for the multiplication of some graphical signal, signal propagation is enacted on the said graph. In a formal expression, the propagated signal, denoted as y, is characterized as the outcome of the graphical shift operator multiplied by the given graphical signal x. Owing to the sparse nature of S, the initial component of the distributed signal (y_i) is exclusively influenced by, and indicated as, x_j for all j within the neighborhood of i. These correspond to the values of the signal (x),

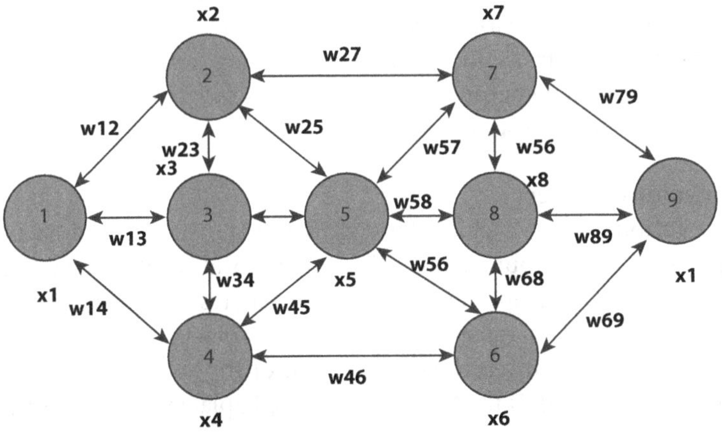

Figure 5.1 Graph with nine nodes.

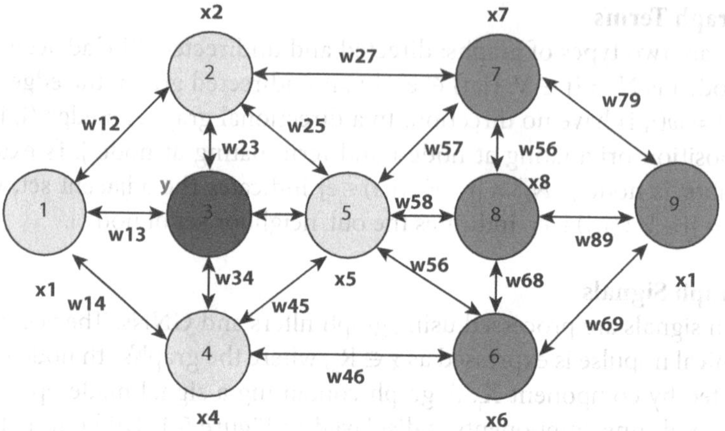

Figure 5.2 Diffusion operation on the graph.

which is the input, that can be processed at the adjacent nodes (j), representing the neighbors of i.

An input graph signal with xi components is shown in Figure 5.2, and node 3's input value is highlighted. Only the levels of the input signal xj at the surrounding nodes, which we show in yellow, have an impact on the outcome of the diffusion operation at this node. The input value x4 has an impact on the diffusion value y3, but x8 has no impact on it.

5.3 Graph Convolutional Filters

Graph convolutional filters (GCNs): The neural network architecture known as a "GCN" performs convolutional operations on graphs. They are important for tasks like node classification and link inference in graph-structured data because they make it possible to extract characteristics and information from nearby nodes in a network.

A. Convolution Filters

A convolutional filter is characterized by the translation and accumulation process of the input signal. While a temporal displacement suggests a postponement, a graph signal shift necessitates contemplation of the inherent topological arrangement [17]. The displaced signal denotes single-step progression when the complement of the A-matrix is the Graph Shift Operator (GSO). If the GSO graph is the Laplacian of L, the displaced signal is the difference at neighboring junctions $[Lx]_i = \sum_{j\in N_i} a_{ij}(x_i - x_j)$.

Once the signal shift is obtained, the graph convolutional filter is simply a weighted sum of multiple shifted signals.

$$H(x) = \sum_{k=0}^{K} h_k S^k x = H(S)x \tag{5.1}$$

where $H(S) = \sum_{k=0}^{K} h_k S^k$ is the N × N polynomial filtering matrix.

$y_i = h_0 x_i + h_1 [Sx]_i + \ldots + h_K [S^K x]_i$, a representation of the signal values arranged in at most K hops, is the final signal at node i. This arises from the condition that $[S^K]_g \neq 0$ signifies the existence of signal propagation. These signals undergo frequent shifts across the graph. The convolution expression for Equation 5.1 originates from the algebraic expansion of the operation, and the discrete-time equivalent can be viewed as a specific instance on cyclic graphs. In this context, the filter in Equation 5.1 is also referred to as a finite impulse response diagram of order K.

Properties

Graph convolution filters hold these properties:

Property 1—Linear: The convolution is linear in the input signal. For two inputs x_1 and x_2 and a filter H(S),

$$\alpha H(S)x1 + \beta H(S)x2 = H(S)(\alpha x1 + \beta x2), \tag{5.2}$$

where α and β are scalars; meaning that the result of an analog representation of outputs is also an analog representation of inputs.

Property 2—Shift Constance: The convolution of the graph does not change with respect to shifts, i.e., SH(S) = H(S)S. This indicates that we can change the sequence of the filters when using two filters, H1(S) and H2(S), each of which operates on the same graph and input signal x.

$$\mathrm{H_1(S)H_2(S)x = H_2(S)H_1(S)x.} \tag{5.3}$$

Property 3—Permutational Equivariance: Convolutions of a graph are equivalent to permutations in the support of a graph. Specifically, the permutation matrices assets are denoted by the following:

$$\mathrm{P = P \in \{0,\ 1\}^{N\times N:}P1 = 1,\ P\ T1 = 1\}} \tag{5.4}$$

Property 4—Parameter Sharing: All nodes share parameters with each other. For two nodes i, j, the respective response signals are $y_i = h_0 x_i + h_1 [Sx]_i + \ldots + h_K [S^K x]_i$ and $y_j = h_0 x_i + h_1 [Sx]_j + \ldots + h_K [S^K x]_j$, which means that the signal that is shifted by k units has a weight of the parameter h_k.

Property 5—Linear Computational Cost: The computational challenge of graph convolution is on the order of O (K|E| + KN). Each shift of the signal from (k) = $Sz^{(k-1)}$ has an order of complexity O (|E |) and must each be multiplied by a scalar hk, implying an added load of O(N). The cost of getting the intermediate output from z (1) ..., z(k) must be scaled by K since this process must be performed K times.

Spectral Filters

These filters are concluded from the Laplacian matrix or the adjacency of a spectrally decomposed graph. Processes, such as graph signal processing, use spectral filters to examine the frequency domain characteristics of graphs [12, 13, 18].

Fourier Transform Graph

The Discrete Fourier Transform (DFT) serves as a mathematical technique by projecting a time signal onto the eigenvectors of the adjunct matrix of a cyclic graph. The Fourier transform, illustrated in the X signal plot, can be displayed as the Graph Fourier Transform (GFT).

$$\breve{x} = V^{-1} x \tag{5.5}$$

Hence, assuming the S is diagonalizable, the eigen decomposition of the graph translator function $S = V diag\,(\lambda)\, V^{-1}$ has the following eigenvectors and eigenvalues: $V = [V_1, \ldots, V_N]$, and $\lambda = [\lambda_1, \ldots, \lambda_N]$)

Similarly,

$$X = V \tilde{X} \tag{5.6}$$

yields the inverse GFT.

Undirected Graphs: The local signal difference in two connected nodes and j can be calculated as Δx_{ij}: = $\sqrt{a_{ij}}\,(x_i - x_j)$), which is higher when the signal in two strongly connected nodes is substantially different. The local variability at node is responsible for the variability in its surroundings.

Frequency Response

By substituting the eigen expansion S = Vdiag(λ)V^{-1} into the graph convolution filter, the output of the filter is as follows:

$$y = \sum_{K=0}^{K} h_k S^k x = \sum_{K=0}^{K} h_k V\, diag(\lambda^{\odot k}) V^{-1} x \tag{5.7}$$

where $\lambda^{\odot k} \in C^N$: $[\lambda^{\odot k}]_i := \lambda_i^k$. Using then Equation 5.4 and bounding the GFT output $\tilde{y} := V^{-1} y$, we can write the spectral relationship of the filter input and output as follows:

$$\tilde{y} = \sum_{k=0}^{k} h_k diag(\lambda^{\odot k}) \tilde{x} \tag{5.8}$$

B. Miscellaneous Graph Filters

Graph filters implement a frequency output and gain greater granularity as the filter order K escalates. However, with higher order, there arises an escalation in the matrix powers of S k introducing numerical instability and diminishing the efficacy of interpolation and extrapolation. Orthogonal polynomials (like Chebyshev polynomials) can mitigate this issue but demand a substantial number of inputs for implementing the expected filter function. Another constraint of the convolution filter lies in its reliance on the graph spectrum rendering it unamenable to approximation by conventional operators. This section elucidates alternative graph filters devised to tackle these challenges.

a. Rational Graph filters

Rational graph filters implement frequency domain output.

$$h(\lambda) = \left(\sum_{q=0}^{Q} b_q \lambda^q \right) \Big/ \left(1 + \sum_{p=0}^{P} a_p \lambda^p \right) \tag{5.9}$$

This is the ratio of two polynomials of degree P and degree Q. This form provides the same frequency response as a convolution filter but with fewer parameters. This is because polynomials require higher orders to obtain identical approximations, whereas exponential functions offer stronger extrapolation and interpolation features [19].

b. Node Domain Filtering

It is also interesting to develop filters in terms of the nodal domain and potentially beyond spectral duality. The obtained result [y]i at node i is typically computed by a graph filter of order K as a weighted sum of the input signals localized in a neighborhood N (i, K).

$$[y]_i = h_{ii}[x]_i + \sum_{j \in N(i,\kappa)} h_{ii}[\chi]_j \tag{5.10}$$

Here, $\{h_{ij}\}$ represent parameters. These parameters also take into account the underlying graph architecture of the signal $[x]_j$ as observed locally from node i. Initially, we establish the connection between the convolutional filter and the operation, then delve into its expansion to encompass node- and edge-specific variations. The graph filter that varies by node utilizes parameters specific to each node h_{ki} to $[x^{(k)}]_i$ and each $[S^k x]_i$. In other words,

$$[y]_i = h_{0i}[x]_i + \sum_{k=1}^{k} h_{ki}[S^k x]_i \tag{5.11}$$

Using the vector $h_k = [h_{k1}$, with different parameters applied with shift k.., $h_{kN}]$ T, you can write a filter like this:

$$H(x) = \sum_{k=0}^{k} diag(h_k) S^k x \tag{5.12}$$

This heightened flexibility enables the implementation of more versatile operators than convolution filters while upholding a localized approach.

c. Nonlinear Graph Filtering

The limitations of node domain filters have been tackled by nonlinear filters, which offer even greater flexibility compared to graph convolutional filters [20–22]. Expressing the output of the graph filter at node involves gathering the signals at that node into the vector $x^{(K)}_i$ and the inputs into the set $H = \{h = [h0, \ldots, h_K]^T\}$.

$$y_i = f(x_i^{(K)}; H) := h^T x^{(K)}_i, \forall i = 1, \ldots N; \tag{5.13}$$

That is to say, when employing h parameters shared among all nodes, it resembles a regression in $x^{(K)}_i$. Node-specific filtering represents a linear adaptation of Equation 5.13 with separate parameters for each node, as can also be demonstrated in Equations 5.11 and 5.12. On the other hand, nonlinear graph filters take into account a nonlinear function $f\left(x_i^{(k)};H\right)$ with the same parameter H for every node. Although the function f (•) in Equation 5.13 might be anything at all, two models that were influenced by conventional signal processing have received significant attention.

Volterra Filter: An extension of Equation 5.13 is to consider a multi-variable polynomial regression in the variable x_i as follows:

$$y_i = f(x^{(K)}{}_I;\, H) := \text{poly}\,(L_0,\ldots,L_K(x^{(K)}i;\, H), \tag{5.14}$$

Median Filter. All the aforementioned types depend on signal interference over a graph. If a specific node exhibits an anomaly, such as an unusually large signal value, it impacts all neighboring nodes and its response. Median filters are suggested as a strong alternative to address these problems. Consider a non-negative integer (h) and a real number (x).

$$y_i = f\left(X_i^{(K)};H\right) : Med\left(h_0\Diamond[\chi]_i;\ldots;h_k\Diamond[S^K X]_i\right) \tag{5.15}$$

Here, the middle operation Med (•) produces the argument in the middle after sorting the arguments in ascending order. Weighted linear representation of the x inputs is calculated to get $[S_x]_i$. Here, the weight is given by the value of row i. If a node has a special value, it can be strengthened by displacement. The operator (S) is present in nearly every node ($X_i^{(K)}$). The median operator mitigates such effects.

D. Regularization-Based Filtering

As mentioned earlier, graph-based parametric function modeling have input–output relationships. If the spectral properties of this mapping are unknown or if data are scarce, designing these parametric filters can be challenging potentially leading to overfitting. In such scenarios, prior knowledge of specific characteristics of the graph signal can be utilized to perform graph filtering *via* regularization. For simplicity, we will focus on using graph tuning to eliminate noise from the signals.

When a graph signal (y) is derived from an irregular dataset (x = y + n), where (n) represents noise, consider the following approach. This issue can be resolved by the following:

$$f(x,y)+\gamma r(y,G), \tag{5.16}$$

a. Smooth Filtering: These strategies employ regularization to decrease signal discrepancies across neighboring nodes. For non-directional graphs, two widely used techniques are Tikonov regularization [6, 18] and Sobolev regularization [23]. In contrast, for directional graphs, total disparity regularization is commonly utilized.

A noiseless signal y on an unidirectional graph has the lower degree $TV_2(y) = y^T Ly$. While TV2 is a regulator, we can get the following by solving for a smooth noise-free signal:

$$\| x - y \|_2^2 + \gamma yTLy \tag{5.17}$$

Increasing $\gamma >> 0$, makes the solution smoother. The closed-form solution of the quadratic convex problem [46] is

$$y * (I + \gamma L) - 1x \tag{5.18}$$

A rational filter, the Tikhonov filter, has a frequency output $h(\lambda) = (1 + \gamma\lambda) - 1$. Understanding the function of the frequency response parameters is also helpful; the best solution to Equation 5.18 is a small frequency graphical filter and a higher-order low-frequency filter.

Sobolev: increases flexibility through more efficient rational frequency output. More specifically, it focuses on solving.

$$\| x - y \|_2^2 + \gamma y^T (L + \epsilon I)^\beta y, \tag{5.19}$$

with $\epsilon \geq 0$ and $\beta \in R_+$. Closed type solution:

$$y^* = (I + \gamma (L + \epsilon I)^\beta)^{-1} X \tag{5.20}$$

$h(\lambda) = (I + \gamma (L + \epsilon I)^\beta)^{-1}$ corresponds to the rational filter and the frequency response. Here, the scalar β controls the order of expression of this function.

Total Variation: When the graph is oriented, we calculate the variable as the difference of the signal and the shape of the change Sy. Using the principle of total variation [20, 24], we can reconstruct the smooth signal through the directed graph by solving the following:

$$||x-y||_2^2 + \lambda ||y - Sy||_1 \tag{5.21}$$

There is an equation as follows:

$$y^* = (I + \gamma(I\text{-}S - S^T + S^T S))^{-1} x \tag{5.22}$$

This is also a reverse filter.

b. Sparsity Filter: These approaches use *a priori* information that signals have discontinuities between adjacent signals, for example, a patchy smooth signal may have uniform outputs inside a bunch of nodes, but have huge arbitrary differences. The sparse filter is used by the graph trend filter (GTF), while otherwise, it is executed by the total disparity.

Suppose $\Delta \in R^{N \times |E|}$ is a directed incidence matrix in an undirected graph G. Sure! Here is the sentence with one randomly selected synonym for each word:

The lines of a bidirectional network G are marked by vertices and the fields by connections. That is, if an edge $e_l = (i, j) \in E$ is mapping node i to node j (i < j), then the entries of column l Δ are as follows:

$$[\Delta]_{ml} = \{-1, \quad m = j,\ 1, \quad m = i,\ 0, \quad otherwise.$$

The action $\Delta^T x$ computes the pairwise variance between the signal magnitudes of every edge. Thus, ΔT might be construed as a chart discrepancy operator. In fact, as L = ΔΔT, the controller in Equation 5.16 may be expressed as the squared form of the l2-magnitude discrepancy vector. Conversely, GTF addresses standard layout issues.

$$||x-y||_2^2 + \gamma ||\Delta^T y||_1 \tag{5.23}$$

This calculates the absolute difference in the output fluctuations of the connected nodes. Problem 5.23 estimates a y-signal of order K = 1 GTF such that the difference is non-zero at only a small number of edges. Higher-order GTFs replace the incidence matrix Δ(1): = Δ in Equation 5.23 with a higher-order version K ≥1.

$$\Delta^{(K+1),T} = \begin{cases} \Delta\Delta^{K} = L^{\frac{K+1}{2}} \; for\ odd\ K \; \Delta^{T}\Delta^{(K),T}, for\ even\ K \end{cases}$$

Wiener Filtering: The high-tuning Wiener filter does not take into account the statistical behavior of the real signal. If this signal exhibits broadly constant graph behavior [24, 25] or is generated by a Markov Gaussian random field [26], the signal and noise statistics can be incorporated into the Wiener filter concept for optimal signal recovery.

First, take an impulse of the order $y \sim D(0, \Sigma_y)$ with covariance matrix Σy and assume that the high-frequency unwanted signals are additive with zero mean and covariance matrix Σn. Assuming both are independent of each other, the Wiener filter involves solving the following:

$$\| x - y \|_2^2 + \lambda \| y - Sy \|_1 \tag{5.24}$$

Then, put the equation on y = H*x.

$$h(\lambda) = \frac{\sigma_d^2(\lambda)}{\sigma_d^2(\lambda) + \sigma_n^2(\lambda)} = \frac{1}{1 + \frac{\sigma_n^2}{\sigma_d^2}}, \tag{5.25}$$

It is a number, and its output at pitch λ is governed by the SNR^{-1} (λ): = $\sigma^2_n(\lambda)/\sigma^2_d(\lambda)$. By contrasting Equation 5.25 with alternate adjusting filters, we observed that the Wiener strainer does not necessitate a set adjusting mass γ for every frequency λ, but rather a frequency-adjusting tuning provided by the reciprocal SNR. Resembling the rationale graph filter talked about earlier, the result of the strainer is acquired by employing the gradient technique. The matrix Σ_n ~~W~~ Σ_n is typically packed. The primary hurdle of these strainers is to recognize a suitable overseer or fusion of overseers to depict the information. This is frequently a tough assignment that demands field know-how; thus, a straightforward resolution employing broader graph strainers as input/output mappings is suggested.

E. Multi GSO Filter

Usually, shifting operations are performed with time delays, and graph settings often allow different choices of graph transformation operators on the data, especially in abstract networks. Both filter coefficients and GSO are difficult to design or learn due to the powers of S that appear in the filter equation and the high degrees of freedom (DoF) that may result in overfitting. The answer to these problems is to construct graph filters that work with multiple predefined GSOs [27, 28]. If Q GSO $\{S_q\}_{q=1}^{Q}$ is given, then, the filter is defined as follows:

$$H(x) = \sum_{q=1}^{Q} \sum_{k=0}^{K} h_{qk} S_q^k x = H(\{S_q\}_{q=1}^{Q})x \tag{5.26}$$

The first discussion of multi-GSO graph filtering uses the combination of two random walk matrices on a directed graph. We also discuss the case of Q = 2. This scheme proposes a filter that enhances group dispersion of the form:

$$\{h_1^*, h_2^*\} = \frac{1}{2\mu} \| e \|_2^2 + \frac{\| h_1 \|_2^2}{2\alpha} + \frac{\| h_1 \|_2^2}{2(1-\alpha)} \tag{5.27}$$

$$\mathrm{e} = \mathrm{y} - \mathrm{H}\left(\{S_q, h_q\}_{q=1}^{2}\right)\mathrm{x} \tag{5.28}$$

where {x, y} are input–output pairs, $\mu > 0$ and $\alpha \in]0, 1[$ are scalars, and hq is the vector of filter parameters with respect to GSO square.

5.4 Graph Filter Banks

It is often sufficient to smooth data, identify discontinuities and jumps, and differentiate signals. However, you can also combine the results of various filters to create more refined representations of your data. When the graph signal of interest has prior modeling information that the coefficients of the filter bank belong to a class of signals that exhibit a specific structural pattern, it can be used in regularization problems scattered.

A. Single-Level M Channel M Indecomposable Graph Filter

A uni-level filter bank that is uni-sampled uses M distinct filters to a signal and sums up the results into one array.

$$\alpha := \left[\mathrm{H}_1{}^{(S)}{}_{\mathrm{x}};\ \mathrm{H}_2{}^{(S)}{}_{\mathrm{x}};\ \ldots;\ \mathrm{H}_\mathrm{M}{}^{(S)}{}_{\mathrm{x}}\right] \tag{5.29}$$

If the filter, often called an analysis filter, is monotonous, it forms a linear transformation from X_N (input signal) to X_{MN} (output signal). The same is true for all MN, which is the dot product of the graphical signal x and with the form $\varphi_{im} := H_m(S)\delta_i$. Here, if j = i, $[\delta_i]_j = 1$, else, it is 0. Each atom φ_{im} is defined through a filter $\tilde{h}_m$ in the spectral domain and can be viewed as a pattern centered at vertex i.

The usual way to regenerate the input is to use the synthesis bank. For a nondeterministic signal, the regenerated signal is by the following:

$$X_{rec} = \sum_{m=1}^{M} G_m(S)H_m(S)x, \tag{5.30}$$

where $\{G_m(S)\}$ is the synthesis filter.

Percival Frame. A sufficient condition for a dictionary atom $\{\varphi_{im}\}_{j=1,2,N;n=1,2,M}$ constitutes a strict Parsival frame:

$$\sum_{m=1}^{M} [\tilde{h}_m(\lambda_i)]^2 = 1, \text{ for } I = 1,2,N \tag{5.31}$$

The advantages of designing a filter to satisfy these conditions are as follows: (i) The filter bank conserves signal energy and also helps prevent numerical instability. (ii) The signal is entirely reconstructed by identical filters in the bank as in the filter bank.

$$\sum_{m=1}^{M} G_m(S)H_m(S)x = V\left[\sum_{m=1}^{M} [\tilde{h}_m(\lambda_i)]^2\right]V^{-1}x = x. \tag{5.32}$$

Nevertheless, it is impossible to create a filter bank consisting of polynomial filters such that $P_m = \tilde{h}_m(\lambda_i)^2 = 1$ for all $\lambda \in [\lambda_{min}, \lambda_{max}]$ (optimal condition graph for independent filters). The idea is not to calculate all the eigenvalues; so, a tight frame is guaranteed [29–32].

Spectrum Graph wavelet. As with wavelet filter banks for discrete-time signals, you can choose filters that are extended versions of each other (with greater filter spread) [11], with extensive support for high-pass filters at the higher end of the spectrum. It has become increasingly localized. As a result, for signals that are noiseless or corrugated smooth according to the graph structure, spectral graph wavelet filter bank coefficients are low.

B. Downsampled and Critical Chart Filter Bank

Minus reduction in size, the M-channel graph strainer ensemble undergoes an additional transformation. This suffices for many applications, and the fact that the coefficients of certain signals are negligible can be used in machine learning. In certain contexts, it may be preferable to reduce the size of the output coefficients to retain solely those associated with the nodes of the mth channel group V_m, thereby diminishing the overall storage overhead. If PM m = 1 and $|V_m|$ = N, the strainer ensemble is precisely sampled [33]. By employing a comprehensive synthesis strainer ensemble encompassing sampling, straining, and aggregating the output response, the reconstructed signal is provided by the following:

$$_{\text{rec}} = \sum_{m=1}^{M} G_m(S) R_{vm}^T R_{vm} H_m(S) X, \tag{5.33}$$

Here, R_{Vm} $|V_m| \times N$ is a selection matrix.

C. Multilevel Chart Filter Bank

Conventional multilevel strainer collections employ various tiers of straining and sampling to temporal sequence data or visuals. For instance, in a standard logarithmic wavelet strainer collection, at each tier, an additional strainer collection is implemented to the yield of the low-frequency conduit, which is sampled from the previous tier [34–38]. In classical temporal sequence scrutiny or image handling, the underlying domain arrangement demands cyclical sampling (i.e., any other temporal sample is feasible). The principal contrast, and one predicament in graph setup, is that unless the graph is exceedingly symmetrical, each successive tier of the strainer collection delineates a coarser graph in a manner that conserves lucid correspondences between eigenvectors. It is not evident how. Eigenvectors of displacement operators are employed to strain the graph at a tier, and displacement operators on the coarse graph are employed to strain the sampled signals on the coarse graph.

D. Data-Adapted Transforms/Dictionary Learning

Every component examined so far—graphs, strainers, and reduction sampling groups—can be adjusted to either the particular examined graph cue or an additional group of representative training cues, for instance, the technique [39] for acquiring polynomial strainers that furnish sparse depictions of the training cues and the technique [40] for acquiring strainers that furnish a firm framework, with each resultant strainer sub-band apprehending on average the same quantity of vitality. This technique is notably advantageous when the vitality of a representative cue from the class of interest is focused in a condensed region, as indicated by researchers in the case of fMRI cerebral data.

5.5 Graph Neural Networks

A collection of graph filters and (typically) a pointwise non-linear activation function is present in each layer of the non-linear, layered architecture that makes up a GNN [14, 40]. They usually have better expressiveness than simpler non-linear graph filters.

The graph perceptron, a simple extension of the graph filter, is the fundamental component of a GNN. A graph perceptron can be constructed using any of the filters, and the activation function can take many forms. Cascading graph perceptron generates graph neural networks (GNN). Formally, a GNN Φ: XN → XN consisting of L layers is given as follows.

$$\Phi(\mathrm{x}) = \mathrm{x}_{\mathrm{L}} \text{ where } \mathrm{x}_{\mathrm{l}} = \sigma\left(\mathrm{H}_{\mathrm{l}}(\mathrm{x}_{\mathrm{l-1}})\right), \mathrm{l} = 1, \ldots, \mathrm{L} \tag{5.34}$$

Let x_0 = x. In alternate words, the result of the l = 1 layer, i.e., $x_1 = \sigma(H_1(x_0))$, is formed from the GNN's input data, which is a signal that has been processed by the graph perceptron. The following layer receives the signal x1 as an input, processes it using a different graph perceptron, and outputs $x_2 = \sigma(H_2(x_1))$. GNN's outcome becomes the output of the final layer x.

These filters depend on the inputs of each layer. Group all filter inputs into a set *H*, and let *H* be the training set $T = \{X_i\}_{i=1}^{|T|}$. Estimate in a data-driven way by reducing the task-dependent cost function $J : R^N \to R$:

$$\sum_{x_i \in T} J\,\Phi(x_i) \tag{5.35}$$

The choice of filter form H` determines the general characteristics of GNN. Below, we discuss convolutional filters.

A. Convolutional Neural Network Graph

The most common GNN architecture uses one GCF per layer. That is, replace H` in Equation 5.34 with Equation 5.1. This results in a convolutional neural network graph. GCNN [41–45] can be compactly written as mentioned below [24].

$$\Phi(X) = X_L \text{ where } X_l = \sigma\left(\sum_{k=1}^{K} S^k X_{l-1} H_{lk}\right), \tag{5.36}$$

GCNN is also continuous Lipschitz with variations in the underlying graph support, although the constants are slightly changed. However, it can stably handle information in particularly large GSO values. This is a feat that cannot be achieved by the complexity of the line graph alone. This makes GCNN more suitable for problems where information on the eigenvalues of GSO is important.

GNN structure blueprints are advancing swiftly, with numerous fresh blueprints emerging conveniently each month. While we have merely delineated the most universal among them, the reader is left pondering whether these recent blueprints are convolutional and thus align with the identical depiction as the blueprints mentioned earlier, or whether it is judicious to define what is innately impractical. Convolutional can likewise be articulated interchangeably employing alternative strainer frameworks.

B. Non-Convolutional GNN

Logical diagram strainers generate GNNs [46] that are convolutional yet can accomplish sharper frequency shifts with fewer trainable parameters. Cayleynet and ARMANets serve as two instances. Node-variable diagram strainers yield straightforward GNNs as a mechanism for producing learning frequency content. Edge-displacement diagram strainers similarly yield GNN, with diagram transformers [47] and natural diagram convolutions [48] emerging as two of the most prevalent descriptors. In this context, we proposed an extended array of GNN resolutions utilizing edge-shift strainers as a means to delineate the advantages and constraints of distinct architectures and how to balance parameter sharing and permutation equivariance. Both node- and edge-variable GNNs are not permutation equivalent frameworks. This is especially beneficial in scenarios where nodes can be distinguished; for instance, learning diverse behaviors may be advantageous for nodes in a power grid, where certain nodes depict generators, while others portray consumers.

5.6 Conclusion

Graph filters have an irreplaceable role in multiple data analysis applications, especially when dealing with graph-structured data. They enable us to extract valuable information, uncover patterns, and perform essential tasks such as noise reduction, feature selection, and visualization. In this chapter, we have outlined various types of graph filters and their design features. Their ability to capture and process graph-structured data makes them a valuable asset for extracting insights and enhancing decision making in a wide range of fields. While graph filters have shown great promise, there are still challenges to be addressed. These include scalability issues for large graphs, adapting filters to dynamic graphs, and creating interpretable filter designs. As technology continues to advance and research progresses, we can expect graph filters to play an even more significant role in future data analysis and machine learning endeavors.

References

1. Oppenheim, A.V. and Schafer, R.W., *Discrete-Time Signal Processing*, 3rd ed, Pearson, Upper Saddle River, NJ, 2010.
2. Goodfellow, I., Bengio, Y., Courville, A., *Deep Learning, series Adaptive Computation and Machine Learning*, The MIT Press, Cambridge, MA, 2016.
3. Taubin, G., A signal processing approach to fair surface design, in: *ACM Conf. Comput. Graph. Interactive Techn*, pp. 351–358, 1995.
4. Taubin, G., Zhang, T., Golub, G., Optimal surface smoothing as filter design, in: *Eur. Conf. Comput. Vision*, pp. 283-292, 1996.
5. Smola, J. and Kondor, R., Kernels and regularization on graphs, in: *Conf. Comput. Learning Theory*, pp. 144–158, 2003.
6. Zhou, D. and Schölkopf, B., A regularization framework for learning from graph data, in: *Int. Conf. Mach. Learning*, 2004.
7. Bougleux, S., Elmoataz, A., Melkemi, M., Discrete regularization on weighted graphs for image and mesh filtering, in: *Int. Conf. Scale Space Variational Methods Comput. Vision*, vol. 4485, pp. 128–139, 2007.
8. Zhang, F. and Hancock, E.R., Graph spectral image smoothing using the heat kernel. *Pattern Recognit.*, 41, 11, 3328–3342, 2008.
9. Crovella, M. and Kolaczyk, E., Graph wavelets for spatial traffic analysis, in: *Joint Conf. IEEE Comput. Commun. Soc.*, pp. 1848–1857, 2003.
10. Coifman, R.R. and Maggioni, M., Diffusion wavelets. *Appl. Comput. Harmon. Anal.*, 21, 1, 53–94, 2006.
11. Hammond, D.K., Vandergheynst, P., Gribonval, R., Wavelets on graphs *via* spectral graph theory. *Appl. Comput. Harmon. Anal.*, 30, 2, 129–150, 2011.

12. Sandryhaila, A. and Moura, J.M.F., Discrete signal processing on graphs. *IEEE Trans. Signal Process.*, 61, 7, 1644–1656, 2013.
13. Sandyhaila, A. and Moura, J.M.F., Discrete signal processing on graphs: Frequency analysis. *IEEE Trans. Signal Process.*, 62, 12, 3042–3054, 2014.
14. Gama, F., Isufi, E., Leus, G., Ribeiro, A., Graphs, convolutions, and neural networks: From graph filters to graph neural networks. *IEEE Signal Process Mag.*, 37, 6, 128–138, 2020.
15. Ma, Y., Liu, X., Zhao, T., Liu, Y., Tang, J., Shah, N., A unified view on graph neural networks as graph signal denoising, in: *ACM Int. Conf. Inf. Knowl. Manag.*, pp. 1202–1211, 2021.
16. Zhu, M., Wang, X., Shi, C., Ji, H., Cui, P., Interpreting and unifying graph neural networks with an optimization framework, in: *ACM Web Conf.*, pp. 1215–1226, 2021.
17. Oppenheim, A.V., Willsky, A.S., Nawab, S.H., *Signals and Systems, series Prentice Hall Signal Process*, 2nd ed, Upper Saddle River, NJ, Prentice Hall, 1997.
18. Shuman, D., II, Narang, S.K., Frossard, P., Ortega, A., Vandergheynst, P., The emerging field of signal processing on graphs: Extending high-dimensional data analysis to networks and other irregular domains. *IEEE Signal Process Mag.*, 30, 3, 83–98, 2013.
19. Trefethen, L.N., *Approximation Theory and Approximation Practice, Extended Edition*, Philadelphia, PA, SIAM, 2019.
20. Xiao, Z., Fang, H., Wang, X., Distributed nonlinear polynomial graph filter and its output graph spectrum: Filter analysis and design. *IEEE Trans. Signal Process.*, 69, 1–15, 2021.
21. Segarra, S., Marques, A.G., Arce, G.R., Ribeiro, A., Center-weighted median graph filters, in: *IEEE Global Conf. Signal and Inform. Process.*, pp. 336–340, 2016.
22. Segarra, S., Marques, A.G., Arce, G.R., Ribeiro, A., Design of weighted median graph filters, in: *IEEE Int. Workshop Comp. Adv. Multi-Sensor Adaptive Process.*, pp. 1–5, 2017.
23. Giraldo, J.H., Mahmood, A., Garcia-Garcia, B., Thanou, D., Bouwmans, T., Reconstruction of time-varying graph signals *via* Sobolev smoothness. *IEEE Trans. Signal Inform. Process. Netw.*, 8, 201–214, 2022.
24. Girault, B., Stationary graph signals using an isometric graph translation, in: *Eur. Signal Process. Conf.*, pp. 1516–1520, 2015.
25. Perraudin, N. and Vandergheynst, P., Stationary signal processing on graphs. *IEEE Trans. Signal Process.*, 65, 13, 3462–3477, 2017.
26. Zhang, C., Flo&xxx0213;encio, D., Chou, P.A., *Graph signal processing-a probabilistic framework*, Microsoft Research, Redmond, WA, USA, Tech. Rep. MSR-TR-2015-31, 2015.
27. Sevi, H., Rilling, G., Borgnat, P., Harmonic analysis on directed graphs and applications: From Fourier analysis to wavelets, 2018. *arXiv preprint arXiv:1811.11636*.

28. Emirov, N., Cheng, C., Jiang, J., Sun, Q., Polynomial graph filters of multiple shifts and distributed implementation of inverse filtering. *Sampl. Theory Signal Process. Data Anal.*, 20, 1, 1–39, 2022.
29. Shuman, D., II, Wiesmeyr, C., Holighaus, N., Vandergheynst, P., Spectrum-adapted tight graph wavelet and vertex-frequency frames. *IEEE Trans. Signal Process.*, 63, 16, 4223–4235, 2015.
30. Leonardi, N. and Van De Ville, D., Tight wavelet frames on multislice graphs. *IEEE Trans. Signal Process.*, 61, 13, 3357–3367, 2013.
31. Dong, B., Sparse representation on graphs by tight wavelet frames and applications. *Appl. Comput. Harmon. Anal.*, 42, 3, 452–479, 2017.
32. Göbel, F., Blanchard, G., von Luxburg, U., Construction of tight frames on graphs and application to denoising, in: *Handbook of Big Data Analytics*, pp. 503–522, Springer, Cham, Switzerland, 2018.
33. Narang, S.K. and Ortega, A., Local two-channel critically sampled filterbanks on graphs, in: *IEEE Int. Conf. Image Process.*, pp. 333–336, 2010.
34. Pavez, E., Girault, B., Ortega, A., Chou, P.A., *Two-channel filter banks on arbitrary graphs with positive semi-definite variation operators*, Ithaca, NY, USA, 2022, *arXiv preprint arXiv:2203.02858*.
35. Narang, S.K. and Ortega, A., Perfect reconstruction two-channel wavelet filterbanks for graph-structured data. *IEEE Trans. Signal Process.*, 60, 6, 2786–2799, 2012.
36. Ekambaram, V.N., Fanti, G.C., Ayazifar, B., Ramchandran, K., Circulant structures and graph signal processing, in: *IEEE Int. Conf. Image Process.*, 2013.
37. Ekambaram, V.N., Fanti, G., Ayazifar, B., Ramchandran, K., Critically sampled perfect-reconstruction spline-wavelet filterbanks for graph signals, in: *IEEE Glob. Conf. Signal and Inform. Process.*, pp. 475–478, 2013.
38. Sakiyama, A., Watanabe, K., Tanaka, Y., Ortega, A., Two-channel critically sampled graph filter banks with spectral domain sampling. *IEEE Trans. Signal Process.*, 67, 6, 1447–1460, 2019.
39. Thanou, D., Shuman, D., II, Frossard, P., Learning parametric dictionaries for signals on graphs. *IEEE Trans. Signal Process.*, 62, 15, 3849–3862, 2014.
40. Behjat, H., Richter, U., Van De Ville, D., Sörnmo, L., Signal-adapted tight frames on graphs. *IEEE Trans. Signal Process.*, 64, 22, 6017–6029, 2016.
41. Isufi, E., Gama, F., Shuman, D., II, Segarra, S., Graph Filters for Signal Processing and Machine Learning on Graphs, 2022. *arXiv:2211.08854v1*.
42. Vapnik, V.N., Principles of risk minimization for learning theory, in: *Conf. Neural Inform. Process. Syst.*, pp. 831–838, 1991.
43. Defferrard, M., Bresson, X., Vandergheynst, P., Convolutional neural networks on graphs with fast localized spectral filtering, in: *Conf. Neural Inform. Process. Syst.*, pp. 3844–3858, 2016.
44. Gama, F., Marques, A.G., Leus, G., Ribeiro, A., Convolutional Neural Network Architectures for Signals Supported on Graphs, in: *IEEE Trans. Signal Process.*, vol. 67, pp. 1034–1049, 2018.

45. Bruna, J., Zaremba, W., Szlam, A., LeCun, Y., Spectral networks and deep locally connected networks on graphs, in: *Int. Conf. Learning Representations*, pp. 1–14, 2014.
46. Gama, F., Anderson, B.G., Sojoudi, S., *Node-variant graph filters in graph neural networks*, 2021, *arXiv preprint arXiv:2106.00089.*
47. Cucurull, G., Casanova, A., Romero, A., Liò, P., Bengio, Y., Graph attention networks, in: *Int. Conf. Learning Representations*, pp. 1–12, 2018.
48. D. Pim, T.S. Cohen, M. Welling, Natural graph convolutions. US Patent App. 17/239, 580, 2021.

6

GCNN with Adaptive Filters for Hyperspectral Image Classification

U. Moulali[1]*, R. Vijayarangan[2], S. Khaleel Ahamed[1] and Kamakshaiah Kolli[3]

[1]CSE Dept, Methodist College of Engineering and Technology, Hyderabad, India
[2]Department of Computer Science and Engineering, Sengunthar Engineering College, Tiruchengode, Tamil Nadu, India
[3]Department of CSE, Geethanjalai College of Engineering and Technology (Autonomous), Nellore, Andhra Pradesh, India

Abstract

Hyperspectral image classification has been used in a wide range of applications such as environmental monitoring, agriculture, and defense. Graph convolutional neural networks (GCNNs) have rapidly become one of the most powerful techniques for modeling graph structured data and open up an entire new field in deep learning. This paper offers a comprehensive introduction to the filtering operations developed over graphs within GCNN literature. At the very center of GCNNs are graph filtering operations that play a vital role in processing, learning, and prediction on graphs. These operations are organized into spatial domain filtering, spectral domain filtering, frequency domain filtering, time domain filtering, hybrid filtering and attention mechanisms, graph pooling (DGCNN), and non-local filtering. In each title, different key methodologies are elaborated such as Graph Convolutional Networks (GCN), ChebNet, Graph Wavelets), Graph Attention Networks (GAT), and Non-Local Graph Neural Network (NL-GNN). The paper discusses practical challenges (over-smoothing, generalizability to different graph structures and unseen graphs, efficiency for large-scale graphs, and interpretability) of GCNNs. The overview illustrates this diversity of techniques that can be used for analyzing and learning from graph-structured data while shedding some light on these challenges. Knowledge of the basics of graph filtering explained in this article is an essential prerequisite for some cutting-edge applications like social network analysis, recommendation systems, drug discovery, and

**Corresponding author*: moulali.u@gmail.com

Malini Alagarsamy, Rajesh Kumar Dhanaraj, J. Felicia Lilian, Vandana Sharma and Gheorghita Ghinea (eds.) Graph Convolutional Neural Networks for Computer Vision, (117–140) © 2026 Scrivener Publishing LLC

image segmentation. Next-generation GCNNs will have to overcome these challenges and improve their model interpretability if they are ever going to be useful in other domains outside graph-based ones.

Keywords: Graph convolutional neural networks (GCNN), graph filtering operations, spatial domain filtering, spectral domain filtering, attention mechanisms, non-local filtering

6.1 Introduction

The GCNN, as a new application for hyperspectral image classification between remote sensing and machine learning [1], is advanced in recent years.

The GraphSAGE model introduced in [2] significantly enhances GCNN scalability through inductive learning. Graph Convolutional Neural Networks (GCNNs) have emerged as a breakthrough architecture in machine learning field [3], which enables the execution of an effective data analysis on graph-structured data. Almost everywhere, we use graphs in the present world. These are commonly used to illustrate complex relationships, networks, and dependencies of data across a range of domains such as social networks, biology, economics, recommendation systems, etc. [4]. GCNNs have emerged as a revolutionary framework that provides us with the ability to take advantage of this sophisticated network of interrelated information. One of the key components of GCNN depends on graph filtering operations. These operations are important to enable information

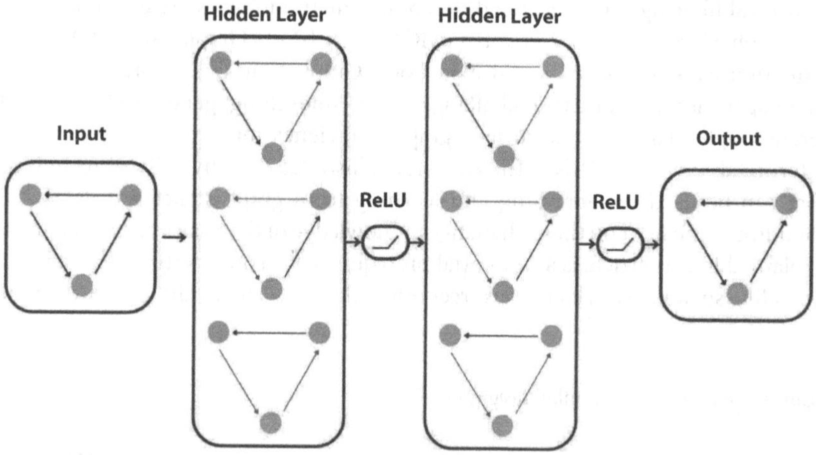

Figure 6.1 Architecture of GCNN.

transmission, feature extraction, and predicting modeling within the network. Architecture of GCNN: As illustrated in Figure 6.1.

In this systematic analysis, our target is to filter out the complex query design and work within detailed categorization of graph filtering operations with the granularity of operation–design–method–operations, so that one can have a clear insight in how these are organized [5]. The filtering operations reveal information about the plethora of procedures as well as methods that aid in the execution of GCNNs. These exist to create what GCNNs operate upon.

We should start by setting up the scenario and provide proper background of why GCNNs are a fundamental structure and for what reasons did they come into existence. Convolutional Neural Networks (CNNs) permanently changed the landscape of Machine Learning, particularly for tasks built upon an organized grid-like input data like image and audio recognition. In particular, CNNs [6] have achieved spectacular success in image recognition. However, they have very limited capabilities when exposed to non-Euclidean or heterogenous data such as graphs. This eliminates the possibility of graphs from working for them. This is how GCNNs came to be—where graph data were difficult to train on in previous generations. We need specialized computers that can both understand the kind of data they will encounter and learn from it.

At its core, GCNNs bring the convolutional operation to graphs as a natural gradient of the classic discrete case of classical regular grids and introduce a new way to process information. In addition, those networks are able to represent the intrinsic structural properties of graphs but also adaptable to the diverse forms that data inside those structures take. The graph filters that operate on nodes, gather information from nearby nodes, and learn from the representations generated by these operations are critical to this adaptation. To effectively leverage the strength of GCNNs, a comprehensive view in terms of different types and approaches involving within these graph filtering operations are recognized to be very essential.

The types of graph filtering operations are as diverse as the graph data itself. The elements of this taxonomy include spatial domain filtering, spectral domain filtering, frequency domain filtering, time domain filtering/hybrid convolution layers, attention mechanisms (Chapter The transversal nature of graph-structuring data are revealed by the multiplicity of methodology, approaches, and possible applications that are exposed in each section).

Also, the success of GCNN is varied with several important matters and factors to be considered. Such concerns include the risks of over-smoothing, as well as making models that can adapt to different graph layouts or generalize to novel graphs unseen at training time and complex computations; a related issue is search for prior knowledge and interpretability in model outputs.

In summary, in this article, through an extensive study on the methods of filtering operations in GCNNs since their birth and until today, we tried to dig into the deep structure of techniques that have empowered us to learn from—and predict with—graphs. These techniques are well described in this investigation line. This level of sophistication is needed to enable creative application in nearly all domains. This insight has important implications for domains including social network analysis, recommendation systems, bioinformatics, and image segmentation. Solving the current limitations and enhancing interpretability become critical to preserve innovation in industry-standard GCNNs as they keep evolving.

6.2 Related Works

Graph Convolutional Neural Networks (GCNNs) and the associated graph filtering mechanisms have been the focus of a significant amount of research effort. In the second part, we will go through a few studies that are key to understanding of GCNNs based on known works on graph filtering operations and development of this field.

Semi-Supervised Classification with Graph Convolutional Networks (GCN)— uses Graph Convolutional Networks to classify data. In this seminal paper, the authors first introduced the Graph Convolutional Network (GCN), which can be considered a base work in graph convolution network and has an important status in heterogeneous GCNN arena. It provides a semi-supervised learning model based on the graph, which can categorize nodes in the network by adding a one-step node neighborhood convolution layer to train suitably and efficiently. Afterward, the GCN architecture was used as a basis for subsequent refinements to the GCNN.

Scalable inductive learning of representations on graphs revolves around learning distributed node representations over complex networks and has liaisons to myriad domains. The GraphSAGE framework is the foundation for the development of methods to enable GCNNs to generalize its predictions to unobserved data.

In image classification tasks, the use of adaptive filters helps in the increase of accuracy and robustness for a great number of scenarios. An adaptive filter is the type of filter that does not have fixed or non-adaptive parameters, which just adjust their values simply based on the input data. For an image classification task, each of these filters will be able to "learn" to automatically adjust itself based on the variations and intricacies found in different parts of an image, hence making it more capable for feature extraction from the input.

The other specific case of using adaptive filter in image classification is for hyperspectral images.

Feature-Based Classification: Due to the large amount of spectral information in hyperspectral imagery, traditional classification techniques often fail. Adaptive filter implementations also improve the filter, as this allows the CNN to localize specific spectral signatures of scene materials and enhance classification discrimination. This adaptability plays an important role in situations when the spectral properties change from one part to another place on the same image.

Moreover, the adaptive filters help to overcome problems related to noise and lighting conditions changes. These filters, by adapting their parameters, can remove noise and amplify useful features to make image classification robust against such perturbations.

To sum up, if we want to solve a real-world hyperspectral image classification problem based on the supervised learning scheme essentially with the prey of adaptive filters, the results above show that they can adapt themselves mainly to face diverse scenarios into it (i.e., specialized local adaptations). This adaptability of (deep) neural networks led to the rise of many excellent image classification models, all taking inspiration from or directly working on Convolutional Neural Networks and because how flexible this architecture is, it quickly became invaluable for basically any task that requires per-pixel decision making—even outside the computer vision realm from environmental monitoring to medical imaging.

Spectral networks (which means networks connected locally to them). In this study, we investigate the spectral domain filtering process of GCNNs. Specifically, the novel trick it illustrates is a method called ChebNet (Chebyshev Polynomial Filtering) to carry out efficient graph convolutions in the spectral domain. Since then, ChebNet has played a significant role in the spectral GCNN architecture development.

Geometric Deep Learning on Graphs and Manifolds provides the theory to understand the geometric aspects of graph structural data; further, it establishes a fundamental aspect in introducing conversion connection S1 spectral and spatial domains of filtering over a graph. In addition, it provides a theoretical side to work on storage and querying data using graphs. This is significant when attempting to formulate strategies for data filtering using spectral and spatial dimensions.

Graph Attention Networks (GATs)—are another major step forward in the realm of GCNNs and colossal improvement on mechanism-based attentions. This paper focuses on self-attention in graph data. Self-attention helps nodes decide how important their neighbors are. Since then, GATs have been broadly used in graph-based applications.

The Graph U-Nets are an important step forward for regular global pooling operations over graph data. Results: We show the effectiveness of the proposed graph U-Net variants in hierarchical feature extraction and demonstrate their usage in various applications such as graph featurization, classification, and segmentation. This work was done by authors of the previous paper in particular.

Spectral Graph Network (GAT)—achieves an encouraging hybrid solution by blending ideas from the spatial and spectral filtering regimes. It provides tools for you to do this using procedures known as summation and concatenation, which gather up local or global data. Hence, GIN has outperformed over a range of benchmark datasets.

In summary, an alternative non-local graph filtering mechanism is the Non-Local Graph Neural Network (NL-GNN) by [7]. In this investigation, a non-local operation is introduced to encode global information into the graph able to improve community detection and classification applications.

Graph signal processing alongside some application case studies and challenges is introduced in [8].

In such a great review paper, we see an overall concept of graph signal processing, including utilization of the graph filtering processes and others. It is useful for situating the broader spectrum of techniques and methodologies in graph data analysis.

Both this seminal work, and a great many others, have significantly shaped the development of GCNNs and graph filtering tools in general. Chances are high that their works will have large impacts on the landscape of techniques, approaches, and methodologies implicating ways for GCNN to be used in different domains and dealing with the main open challenges such as over-smoothing problems, generalization issues, computational complexity-related questions, or interpretability [9]. This interplay of those contributions keeps on driving GCNN design space and the potential for radical applications in these networks.

6.3 Classification of Graph Filtering Operations

The recent years have witnessed much interest in Graph Convolutional Neural Networks, which are capable of performing tasks on the graph-structured data like the social networks, citation networks, and molecular graphs. Graph filtering processes are at the core of how GCNNs process and learn from graph input, so it is not surprising to observe such a significant measure of their performance as it exists in connection with this specific operational detail.

In this comprehensive survey, we will study the taxonomy of graph filtering operations in GCNNs and inspect a wide range of algorithms employed in GCNNs.

6.3.1 Introduction to Graph Convolutional Neural Networks (GCNNs)

GCNNs are a particular kind of neural network that was devised to work on graph structured data. Unlike traditional Convolutional Neural Networks (CNNs), which are designed to handle a grid of data (images), General Convolutional Neural Networks(GCNNs) are formulated for non-euclidean or irregular data. The main challenge is to leverage the structure and information in graph, as a way to enable tasks like feature extraction and classification.

As mentioned above, GCNNs mainly intend to generalize convolution on normal grids (e.g., images) so that it can be used with graph-structured data. The difference is that the filters learned by GCNNs are applied on a node and its neighborhood nodes, rather than convolution to predefined kernels that traditional CNNs do. Most activities with graph filters are all about these filter repos.

6.3.2 Classification of Graph Filtering Operations

Adaptive filters are used in hyperspectral image classification where they adjust their filter coefficients dynamically and adaptively to improve the extraction of useful features, thus leading to a more accurate overall classification. Hyperspectral image adaptive filter: Adaptive filters based on the graph filtering operation may be used for hyperspectral data along with the classification and thereby allows to adaptively process and filter information according to specific spectral characteristics present anywhere in an image region.

Hyperspectral image classification (stage 2): An adaptive filter, namely, selective support vector machine, addressing the constraints due to variations in spectral signatures, noise, and illumination conditions. By automatically adjusting the filter parameters, adaptive filters can efficiently suppress noise and emphasize spectral features of interest to enhance representation of underlying information in hyperspectral data.

Moreover, methods, such as adaptive graph convolutional networks (AGCNs), can perhaps be developed for the integration of adaptive filters in hyperspectral image classification under the context of graph filtering operations. It proposed a gating mechanism, specifically designed to operate in the spectral domain, as its main difference with previous works aiming at generalizing convolutional filters: different materials exhibit different spectral response patterns (and no invariant mapping from wavelength space to spatial domain exists), and thus, it can be beneficial that the convolutional layer is able to adapt this pattern according to the local context. Such adaptability increases the model's discrimination ability of classes with little spectral difference, which results in improved classification accuracy.

Furthermore, hyperspectral image classification with adaptive filtering can be combined with graph attention mechanisms. These mechanisms effectively guide the model to concentrate on certain bands or regions of interest during feature extraction and classification.

This means adaptation of filter mask parameters to capture even a little portion of these subtle spectral information present in different locations, like scene and target areas, as optimally as possible in hyperspectral image classification scenarios. Complementarily, it will be even more effective when a graph filtering operation is considered. The adaptiveness of these filters and its better performance on classification are the reasons why they can be a worthy part itself within the changing world of graph-based techniques for hyperspectral data processing.

Birkhoff decomposition-based composition GNNs: This is specifically referring to the filtering process at each node, *via* the boosting of GCN-like configurations or some variants. These various categories are worth a closer look:

6.3.2.1 *Spatial Domain Filtering*

Spatial domain filtering involves directly applying filters to the graph's nodes and their immediate neighbors. This category includes operations like the following:

a. Graph Convolutional Layer

It is one of the fundamental elements in GCNNs. In a Graph Convolutional Layer, each one of the graph nodes has to gather information from its neighbors, pass this neighborhood information through a learnable filter, and propagate it forward for the new layer. Some representative examples include the Graph Convolutional Network, which is also usually simply referred to as GCN, and GraphSAGE. Layers in the Graph Convolutional Network are depicted in Figure 6.2.

Node Selection and Feature Aggregation

Now, consider a node, which is a middle (v-i). This node additionally has connections with its adjacent nodes, and in this step, we gather personal information from these neighbors.

A spatial convolution is done over each of the v-nearby i's nodes, v-j, to gather information about around area. The corresponding operation in a general GCNN is expressed as the following Script to Mime Types:

$$h_j^{(l)} = \sum\nolimits_{v_k \in N(v_i)}^{l} x w_{jk}^{(l)} \cdot h_k^{(l-1)} \tag{6.1}$$

where:

- $h_j^{(l)}$ is the output feature of node v_j in layer l.
- $h_k^{(l-1)}$ is the feature of neighboring node v_k in the previous layer $l-1$.

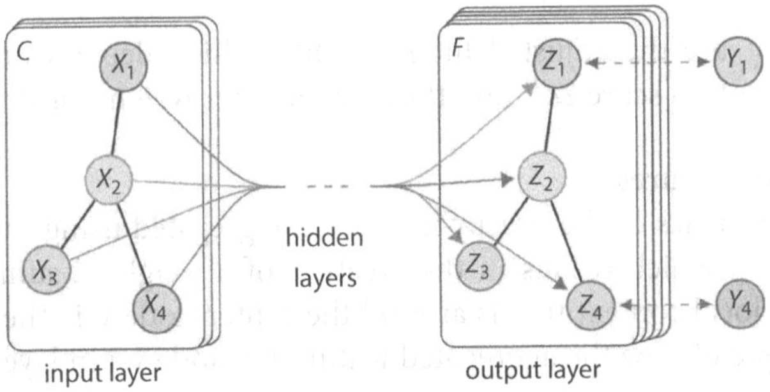

Figure 6.2 Layers used in graph convolutional network.

- $N(v_i)$ represents the set of neighboring nodes of v_i.
- $w_{jk}^{(l)}$ are fixed filter weights for the convolution operation in layer l.

Adaptive Filter Weight Calculation

In GAM-AFW, we introduce an attention mechanism through which the coefficients of filters can be dynamically adjusted according to the importance of nodes around it. The attention mechanism calculates the so-called attention for each neighbor, which specifies how much each of them should be taken into consideration. The attention mechanism might be represented as follows:

$$\alpha_{ij}^{(l)} = softmax\left(f\left(h_i^{(l-1)}, h_j^{(l-1)}\right)\right) \tag{6.2}$$

where:

- $\alpha_{ij}^{(l)}$ represents the attention score for the connection between v_i and v_j in layer l.
- $f(\cdot)$ is a learnable function that computes the compatibility score between nodes v_i and v_j. It can be implemented as a neural network layer.

Adaptive Filter Weight Adjustment

After that, the attention scores that were received from the attention mechanism are employed to make dynamic adjustments to the filter weights. The weights of the adaptive filter may be expressed in the following way:

$$\tilde{w}_{jk}^{(l)} = \alpha_{ij}^{(l)} \cdot w_{jk}^{(l)} \tag{6.3}$$

where:

- $\tilde{w}_{jk}^{(l)}$ are the adjusted filter weights, which depend on the attention score $\alpha_{ij}^{(l)}$ and the original filter weights $w_{jk}^{(l)}$.

Aggregated Features

The characteristics of the nearby nodes are aggregated using the adaptive filter weights, which results in the creation of a weighted sum that represents the local context that is around the center node v-i. The following is an example of how the aggregated feature for node v-i in layer l may be represented:

$$h_i^{(l)} = \sum_{v_j \in N(v_i)} \tilde{w}_{ji}^{(l)} \cdot h_j^{(l-1)} \tag{6.4}$$

Node Embedding Aggregation Equation

For each node v_i and its sampled neighbors $N(v_i)$, GraphSAGE defines the aggregation operation as follows:

$$h_{agg}(v_i) = AGGREGATE\left(\{h(v_i)\} \cup \{h(v_j) \forall v_j \in N(v_i)\}\right) \tag{6.5}$$

Here:

$h_{agg}(v_i)$ represents the aggregated embedding for node v_i.
$h(v_i)$ is the embedding of the central node.
$h(v_j)$ are the embeddings of the sampled neighboring nodes.

- $AGGREGATE(\cdot)$ is an aggregation function (e.g., mean pooling, LSTM, or pooling operation) that combines the embeddings of the central node and its neighbors. The choice of aggregation function depends on the specific implementation.

Sampling Strategy

To guarantee that its computations are as efficient as possible, GraphSAGE employs a sampling technique that chooses a predetermined number of neighbors for each node. As can be seen in the equation above, the phase of aggregation incorporates the neighbors who were randomly picked. The method of sampling could comprise random walks, uniform sampling, or any number of different approaches that strike a healthy balance between exploration and exploitation.

Objective Function

GraphSAGE is often trained to minimize a loss function that assesses the similarity between the learned embeddings and the ground truth labels or targets.

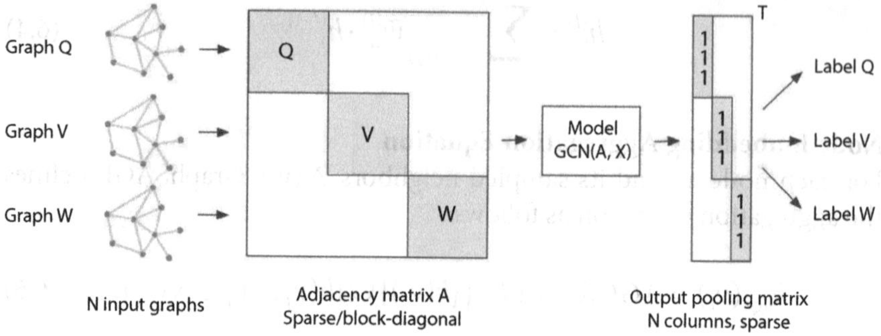

Figure 6.3 Block diagram of the GraphSAGE approach.

Figure 6.3 illustrates the GraphSAGE approach using adjacency matrices and pooling mechanisms.

Graph Laplacian

In ChebNet, the graph Laplacian matrix L is used as the basis for spectral filtering. The graph Laplacian is defined as $L = I - D^{-1/2}AD^{-1/2}$, where:

- A is the adjacency matrix of the graph.
- D is the degree matrix of the graph.
- I is the identity matrix.

ChebNet Spectral Filtering

The spectral filtering operation in ChebNet can be approximated as follows:

$$f(L) \approx \sum_{k=0}^{K-1} \theta_k T_k(L^{\sim})X \tag{6.6}$$

where:

- $f(L)$ represents the filtered signal using Chebyshev polynomial approximations.
- θ_k is the Chebyshev coefficient to be learned during training.
- $T_k(L^{\sim})$ denotes the k-th Chebyshev polynomial of the rescaled Laplacian $L^{\sim} = 2L/\lambda_{max} - I$, where λ_{max} is the largest eigenvalue of L.
- X is the input feature matrix, where each row corresponds to a node and each column to a feature.

Graph Fourier Transform Equation
The GFT of a graph signal x can be expressed as follows:

$$\hat{x} = \Phi^T x \tag{6.7}$$

where:

- $\hat{x}$ is the graph signal in the GFT domain.
- Φ is the matrix of eigenvectors of the graph Laplacian, where each column ϕ_i is a basis function associated with an eigenvalue λ_i.
- x is the original graph signal.

Inverse Graph Fourier Transform Equation
To recover the original signal from its GFT, the inverse GFT can be calculated as follows:

$$x = \Phi \hat{x} \tag{6.8}$$

6.3.2.2 *Frequency Domain Filtering*

Frequency domain filtering methods operate on graphs by considering the frequency components of signals on the graph. Frequency domain filtering in the context of graph signal processing is a novel method that extends the concept of filtering from traditional signal processing to graph structured data. It involves transforming the graph signal into a frequency domain, applying filters, and then transforming it back to the original domain. Below, I will describe the method with equations. They are inspired by signal processing techniques and include the following:

1. Graph Fourier Transform

- The first step in frequency domain filtering is to transform the graph signal into the frequency domain using the Graph Fourier Transform (GFT). As mentioned earlier, the GFT represents the signal as a linear combination of graph Fourier basis functions.
- Given a graph signal x and its GFT representation $\hat{x}$, the GFT is expressed as follows:

$$\hat{x} = \Phi^T x \tag{6.9}$$

where:

- $\hat{x}$ is the graph signal in the frequency domain.
- Φ is the matrix of eigenvectors of the graph Laplacian.
- x is the original graph signal.

2. Frequency Domain Filtering

- In the frequency domain, we can apply filters to modify the spectral components of the graph signal. Filters are represented as diagonal matrices in the frequency domain.
- Let us denote the filter applied in the frequency domain as $H(\lambda)$, where λ is the eigenvalue associated with each basis function.
- The filtered signal in the frequency domain is computed as follows:

$$\hat{y} = H(\Lambda)\hat{x} \tag{6.10}$$

where:

- $\hat{y}$ is the filtered graph signal in the frequency domain.
- $H(\Lambda)$ is a diagonal matrix containing the filter response for each eigenvalue λ.

3. Inverse Graph Fourier Transform

- After applying the frequency domain filter, we can obtain the filtered signal in the original graph domain by performing the inverse Graph Fourier Transform (iGFT).
- The inverse GFT is expressed as follows:

$$y = \Phi\,\hat{y} \tag{6.11}$$

where:

- y is the filtered signal in the original graph domain.
- $\hat{y}$ is the filtered signal in the frequency domain.

Graph signal processing is an emerging field in which frequency domain filtering can help make the process more efficient in some applications, like denoising, feature extraction, and signal analysis on graph structured data. The signal in the frequency domain can be filtered such that specific spectral components are ignored to kill particular characteristics or enhance them. The method provides an efficient way of processing and analyzing data on irregular graph structures by extending the key idea of frequency domain analysis from classical signal processing to graphs.

a. Graph Wavelets
Graph Wavelets (scaled graph structure interpreter) present a multi-scale version of the graph and hence make them suitable for any task where relationships exist at different scales.

Graph Signal Processing (GSP)
GSP—Graph Signal Processing is an extension of traditional signal processing concepts to graphs. These tools are used—among other things, some of which may be impossible—to filtering and frequency domain analysis (though link prediction and signal denoising were just two applications we could see).

6.3.2.3 *Time Domain Filtering*

Time domain filtering is a relatively novel approach that maps graph data to sequences. This enables the use of RNN (Recurrent Neural Networks) and temporal filtering.

Graph Fourier Transform: The first step in frequency domain filtering is to transform the graph signal into the frequency domain leveraging the Graph Fourier Transform (GFT). Recall that the GFT discovers a signal as an expansion over graph Fourier basis functions. Given a graph signal x and its GFT representation $\hat{x}$, the GFT is expressed as follows:

$$\hat{x} = \Phi^T x \tag{6.12}$$

where:

- $\hat{x}$ is the graph signal in the frequency domain.
- Φ is the matrix of eigenvectors of the graph Laplacian.
- x is the original graph signal.

3. Frequency Domain Filtering
    - In the frequency domain, we can apply filters to modify the spectral components of the graph signal. Filters are represented as diagonal matrices in the frequency domain.

- Let us denote the filter applied in the frequency domain as $H(\lambda)$, where λ is the eigenvalue associated with each basis function.
- The filtered signal in the frequency domain is computed as follows:

$$\hat{y} = H(\Lambda)\hat{x} \tag{6.13}$$

where:

- $\hat{y}$ is the filtered graph signal in the frequency domain.
- $H(\Lambda)$ is a diagonal matrix containing the filter response for each eigenvalue λ.

4. Inverse Graph Fourier Transform
   - After applying the frequency domain filter, we can obtain the filtered signal in the original graph domain by performing the inverse Graph Fourier Transform (iGFT).
   - The inverse GFT is expressed as follows:

$$y = \Phi\hat{y} \tag{6.14}$$

The methods belonging to this category are outputs.

6.3.2.4 *Graph Neural Networks (GNNs)*

GNNs, e.g., Gated Graph Neural Network (GGNN), treat graph structured data as dynamic sequences. They are based on RNN-like architectures and can process data as they evolve over time; these networks have been employed in recommendation systems and social network response modeling.

6.3.2.5 *Hybrid Filtering*

Hybrid filtering techniques are a composite of spatial as well as spectral mechanisms to make the best of both worlds. All of these methods aim to encapsulate local as well as global information. This integrated method is extremely helpful in instance of graph signal denoising, feature selection, and graph-based classification as a comprehensive understanding of both local and global characteristics is required. Hybrid filtering reflects a combination of different techniques to support a comprehensive approach in graph-structured data analysis, thus providing improved performance across diverse real-world tasks. A hybrid approach would look like this:

a. GIN: Graph Isomorphism Network
GIN combines aspects of spatial and spectral filtering and uses aggregation functions, like sum and concatenation, to maintain situational awareness. Trace: GIN has achieved great performance on benchmark datasets.

6.3.2.6 Attention Mechanisms

In gated CNNs, attention mechanisms are introduced to determine which neighbors the selected filter will interact with. They have become one of the most popular in-network neighbor discovery approaches as they capture high information nodes. Variants include the following:

Graph Attention Networks (GAT)
GAT uses the self-attention mechanism to calculate attention scores for other neighbors so that a node can give more importance to its relevant neighbor [7]. This adaptive filtering process helps the network capture more contexts.

6.3.2.7 Graph Pooling

In graph pooling, the goal is to down-sample a given larger graph by choosing a subset of nodes and summarizing them—i.e., performing aggregation of their node features. Pooling has been applied on GCNNs to reduce the computational cost and get higher representation of graphs as well. Methods include the following:

a. Graph U-Net
Graph U-Net: A pipeline that merges upsampling and pooling layers to natively hold details at the edges of nodes as well as global summaries. It is applicable to tasks such as image segmentation and graph classification [8, 9].

6.3.2.8 Non-Local Filtering

Non-local methods are inspired by non-local means filtering in image processing, which take long-range dependencies into consideration between pixels and form a more precise representation. They are often foundational in how to understand relationships within vast and complicated graphs. Examples include the following:

Non-Local Graph Neural Networks (NL-GNN)

- NL-GNN utilizes non-local operations for graph global context capturing especially for tasks based on graphs like community detection and graph classifications [6].

6.4 Experimental Analysis and Discussion

Graph Convolutional Neural Networks, aka GCNNs, come with different sorts of perils and booming consideration factors, which direct the research pathway alongside the practical implementations. One of the most notorious is over-smoothing—a phenomenon in which too many filter layers within the network hide variations between nodes. A shallow model means that the expressive properties of the given model are maintained between epochs, but doing so comes at a cost to the depth of the model. It has flexibility that can be used to deal with the various graph architectures like directed, bi-partite, or heterogeneous kinds of graphs. To ensure that GCNNs are applicable on any kind of graphs, researchers try to come up with architectures designed not for a specific graph type. Taking into account that inductive learning, .e., the ability to generalize unseen graphs in practice, is still a hard problem to solve, techniques, like GraphSAGE and 38GINraph Isomorphism Networks (GIN), are currently state of the art in trying to enable this sort of generalization. Message passing and computation should be done in an efficient manner when using GCNNs to large-scale graphs. So there is an ongoing search to make model outputs interpretable, as in fields like healthcare where interpretability is critical. Since labeled data are scarce in many real-world graph-based applications, semi-supervised learning approaches, such as transfer learning, have gained more popularity. Concurrently, the landscape regarding GCNN also involves scalability in graph pooling operation, suitability to heterogenous graph anchors and adversarial perturbations, and computational complexity of coarsening. To fully revolutionize the advancement of GCNNs across diverse types of applications and domains, these impediments need to be addressed.

When you deal with big and complicated graphs, the problem of scaling becomes unavoidable, and for Graph Convolutional Neural Networks (GCNNs), this is not different. This is a highly fundamental requirement that should be satisfied. With larger datasets, the computing requirement may become prohibitive. Scientists and industry professionals are seeking methods to down-sample large graphs well, yet efficiently. Due to computational affordability and without significantly decreasing the performance, often, these techniques employ methods like adaptive pooling and hierarchical aggregation as above.

Furthermore, the governance of mixed networks containing also disparate types of nodes and edges is an additional layer to this complexity. There have to be novel ways for GCNNs to adapt to this intricate structure. One way to do so is building metapath-based models. They are designed to capture links and information flow between different types of nodes similarly well. Heterogeneous graph analysis is crucial but has been largely overlooked in fields, like the recommendation system, where unique features exist for people, objects, and interactions alike.

Furthermore, the high level of data noise is a challenge that must be addressed as well as anomaly detection from graph-structured datasets. For GCNNs to be deployed in the real world, they need to robust against noise and have the ability to identify anomalous data points, which can be critical for applications where data quality is of concern. These days, research is being done to assess how noise can be best stripped off, data outliers detected, and models trained in the presence of defective datasets.

In summary, although with serverless we can forget about operating and just focus on building applications, computing resources are still a constraint. Training and deployment of GCNNs can be very computationally expensive; however, so this may present a challenge for smaller companies or academics who do not have access to high-performance computing equipment. The existence of cloud-based services and pre-trained models has played an important role in the democratization of GCNN technology. This has ramped up its use in other industries and domains as well making the tech more widely available and useful.

Answering these varied questions and concerns is necessary to expand the field (GCNNs) and give GCNN real-world appeal and meaning. As ongoing research and innovation drive forward the state of the art in their applications, GCNNs are set to provide an even more pivotal role for insight discovery, prediction, and decision support in complex systems and networks.

Although graph filtering methods present a rich landscape of methodologies to be categorized, there are many challenges and ideas in mind for GCNN design and application, including the following:
Smoothing Too Hard: This happens if the graph is filtered so many times on an edge that there are no differences between nodes. This can be due to over-application of filtering layers. Getting the right level of details while not starving your model is very important.

Graph Structures: The variety of link types in graphs requires specialized filtering processes. However, adapting GCNNs for other types of graphs, such as directed or bipartite graphs, is still an ongoing research issue.

Generalization: It is hard to generalize unseen graphs or do inductive learning. We need to learn models competent of managing changing graph sizes and architectures.

Scalability: When working with large graphs, scalability becomes a big problem to be pursued. To create more practical applications, people need to think first about how to communicate messages and, secondarily, parallelization algorithms.

Interpretable Models: A long running issue has been to understand how GCNNs arrive at their predictions and distill any relevant features from them.

6.5 Conclusion

The environment surrounding graph filtering operations and the applications of Graph Convolutional Neural Networks (GCNNs) are constantly changing and expanding with increasing importance in a multitude of different fields. This comprehensive study has provided a detailed exploration of the vast landscape of graph filtering operation types to uncover numerous strategies that contribute to successful GCNNs in modeling and learning as it pertains to shape data. In general, graph filtering operations are a subset of graph processing operations that remove elements from the graph. To take full advantage of GCNNs in modern applications (e.g., social network analysis, recommendation systems, bioinformatics and image segmentation...), it is necessary to understand how these filtering operations work. There are many ways you can leverage this potential.

Graph filtering operations comprise a rather comprehensive taxonomy of techniques, including spatial, spectral, frequency, and temporal approaches just to cite some examples. The list continues with hybrid methods, attention-based filters (e.g., PPFilters), pooling strategies, or non-local graph filters. There is a different role to play for each category, based on the functions it enables us retrieve its vital features and particulars from a wide range of graph data structures. Fundamental contributions establishing the foundation for wider developments enabling Graph Convolutional Neural Network (GCNN) area like the Graph Convolutional Network (GCN), GraphSAGE, ChebNet, and Graph Attention Networks (GAT). These early contributions laid the foundation for the development of filter methods on graphs and their application in a range of fields.

GCNNs bring a series of issues, including, but not limited to, over-smoothing, adaptation to different graph architectures, generalization in unseen data contexts, and/or efficiency in large-scale graphs or the transparency of the model output. Solving such problems is indispensable for GCNNs to gain long-term success and maintain its usefulness in real-world scenarios. The power of GCNNs to affect graph-structured data is universal, not specific to anyone area. GCNNs have a lot of applications such as on social networks and molecular science among others. They have been found to be indispensable especially in the areas where conventional machine learning methods fail because of their flexibility and generalization ability. The landscape of GCNN environments keeps evolving as can be seen with an ongoing research to enhance the earlier established practices or come up with newer methods. Consequently, work in obtaining interpretable models with better generalization and more efficient computation is a major pathway of research.

In a world of increasingly complex and interlinked data, we expect this role to be even more important for GCNNs and the graph filtering operations they perform. Being able to detect novel patterns, predict future outcomes, and extract valuable knowledge from graph-structured data could potentially disrupt established approaches across a range of sectors including healthcare, social sciences, or finance.

It is through coordination between researchers, practitioners, and the broader scientific community that we will be able to overcome these challenges and advance further on what seems a very promising path provided by this novel technological cohort in our journey toward an ever-evolving field of GCNNs and graph filtering operations. In doing so, we are traversing through the rapidly changing landscape of GCNNs and graph filtering operations. The journey of unlocking the true power behind GCNNs remains ongoing, and there are a multitude of paths ahead as we persist in exploring deeper dimensions of intelligence inside rich structured graph data. The future is alive with possibilities because we are still probing the vast unexplored stereotypes.

References

1. Fu, S., Liu, W., Zhang, K., Zhou, Y., Tao, D., Semisupervised classification by graph p-Laplacian convolutional networks. *Inf. Sci.*, 560, 92–106, 2021, https://doi.org/10.1016/j.ins.2021.01.075.
2. Hamilton, W.L., Traditional Graph Generation Approaches, in: *Graph Representation Learning. Synthesis Lectures on Artificial Intelligence and Machine Learning*, Springer, Cham, 2020, https://doi.org/10.1007/978-3-031-01588-5_8.
3. Bruna, J., *et al.*, Spectral Networks and Locally Connected Networks on Graphs. CoRR abs/1312.6203, p. 16, 2013.
4. Monti, F., *et al.*, Geometric Deep Learning on Graphs and Manifolds Using Mixture Model CNNs, in: *2017 IEEE Conference on Computer Vision and Pattern Recognition (CVPR)*, Honolulu, HI, USA, pp. 5425–5434, 2017, doi: 10.1109/CVPR.2017.576.
5. Su, G., Wang, H., Zhang, Y., Zhang, W., Lin, X., Simple and deep graph attention networks. *Knowledge-Based Syst.*, 293, 111649, 2024, ISSN 0950-7051, https://doi.org/10.1016/j.knosys.2024.111649.
6. Gao, H. and Ji, S., Graph U-Nets. *IEEE Trans. Pattern Anal. Mach. Intell.*, 44, 9, 4948–4960, 1 Sept. 2022, doi: 10.1109/ TPAMI.2021.3081010.
7. Kim, B.-H. and Ye, J.C., Understanding Graph Isomorphism Network for rs-fMRI Functional Connectivity Analysis. *Front. Neurosci.*, 14, 630, 2020, doi: 10.3389/fnins.2020.00630.
8. Liu, M., Wang, Z., Ji, S., Non-Local Graph Neural Networks. *IEEE Trans. Pattern Anal. Mach. Intell.*, 44, 12, 10270–10276, 1 Dec. 2022, doi: 10.1109/ TPAMI.2021.3134200.

9. Ortega, A., Frossard, P., Kovačević, J., Moura, J.M.F., Vandergheynst, P., Graph Signal Processing: Overview, Challenges, and Applications. *Proc. IEEE*, 106, 5, 808–828, May 2018, doi: 10.1109/JPROC.2018.2820126.

7

Graph Convolution Neural Network on Human Motion Prediction

B. Subbulakshmi, M. Nirmala Devi* and Srimadhi J.

Department of Computer Science and Engineering, Thiagarajar College of Engineering, Madurai, India

Abstract
With extreme growth on computer vision, human motion prediction has gained more popularity by predicting motion from a sequence. The evolution in this area of prediction is from predicting the posture of a human to the motion of the human's next step and to predicting human motion with scene context. There are several methodologies and algorithms for the prediction of human motion having better accuracy. Starting from statistical methods, machine learning is employed for prediction combined with neural networks and using ideas of a graph with nodes and edges encapsulated within neural networks to provide good results. In this work, the base idea on graph convolution neural network, which is involved in human motion prediction, is taken as a review by discussing the concept and its application. Graph convolution network applied alone has its limitations; working on those limitations also has extended works using the graph convolution concept. A theoretical review is provided.

Keywords: GCN, CNN, spatiotemporal, motion prediction, GNN, deep learning, RNN, human motion

7.1 Introduction

In this growing AI era, we try to work on automation in all aspects of science. Computer vision is one of the growing research areas where we perform actions, like recognition, detection, and prediction, on images and

**Corresponding author*: nirmaladevi2004@gmail.com

Malini Alagarsamy, Rajesh Kumar Dhanaraj, J. Felicia Lilian, Vandana Sharma and Gheorghita Ghinea (eds.) Graph Convolutional Neural Networks for Computer Vision, (141–160) © 2026 Scrivener Publishing LLC

videos to obtain certain vision-based outcomes, which has a larger impact in applications like healthcare, robotics, security, health diagnosis, which uses images of scan, surveillance, etc. Deep learning helps to work on larger data of images and videos and create various deep learning (DL) and machine learning (ML) models to perform tasks. Vision studies work on various aspects, like natural language processing, recognition, and detection, and predictions work on an object, human, human to human, human to object, object to object, etc. There are many research works on these topics with experimental results, and yet, automation has its defects and errors. Human motion is yet another engrossing topic in computer vision using deep learning models.

7.1.1 Human Motion Prediction

As we all know, prediction means forecasting the future incident. Motion prediction is forecasting the next step of motion from previous motion steps [4]. We are discussing human motion prediction, where human motions, such as walking, running, and sitting, are forecasted from the previous motion action. Why do we need human motion prediction? Here is a simple answer. Consider the scenario of people walking on a street path. Our movement depends on predicting the movements of our fellow people like their movements tend to collide with yours. Forecasting is done in our daily routine. Automating these ideas of prediction in areas of automation helps in applications like healthcare, robotics, automated vehicles, etc. Given a sequence of motions from videos or image frames, prediction is done by learning the past frames analyzing the postures, etc., techniques like employing inputs as human skeleton points, i.e., human skeleton joints are processed as inputs of every movement of joints so that the motion or pose of a human can be detected [5]. With that sequence of skeleton joint points of humans, it is easily possible to predict human motion. Much research contributes to predicting the motion of a human using various techniques, including the skeleton joint approach, which is done and will be discussed in upcoming sections. Motion prediction of humans has become more renowned because of its complexity, like uncertainty in motion, and it is highly dynamic, which gains researchers' attention.

7.1.2 Traditional Methods

Human motion prediction is done using various methods using statistical machine learning algorithms employed for prediction inclusive of various deep learning models for prediction. Algorithms like Hidden Markov's

Models (HMM) and Gaussian process (GPHV) hidden variable models are statistical methods for motion prediction in humans. Markovian assumptions have imposed experts' knowledge about motion in traditional approaches. Hidden Markov Models' state-space equations are used to explore the bandying between inference complication and model capacity. Earlier in the field of animation Hidden Markov Model spaces were used as style parameters for generating human pose sequences. As we all know, the pose of a human can be used for predicting; hence, a pose sequence using the Hidden Markov Model is beneficiary in terms of motion prediction. Even in recent methods, mixtures of Hidden Markov Models for more complex models are used. The limitation of the Hidden Markov Model is that the model may be exact and efficient for learning, but the expressiveness of complex human motions is limited.

To learn temporal dynamics and non-linear motion prediction using expectation exaggeration, Gaussian models or processes are used. [12] describe the Conditional Restricted Boltzmann Machine (CRBM) as a method that models human motion and binary latent space, which needs sampling for inference. [20] and [21] discussed that Principle Component Analysis is a famous machine learning algorithm employed for time-shifted and time-scale activity for human motion models. PCA models are used for pose tracking. Another machine learning algorithm is the Random Forest technique, which is used for predicting the unknown motion of past observations. Another traditional method, Mixture of Experts (MoE) [11], is a machine learning model for predicting human motion. A gating network that combines numerous experts produces a blending coefficient. According to [11], a mixture of experts is a motion classifier that employs a gating network to determine the likelihood that an input motion is included in every class of mobility and fuses the output of pertinent experts to automatically provide the best result. Many human motion models' generalization is much enhanced by this method. Several candidate adjacency matrices are used as experts in Mixture of Experts in addition to employing gating networks to train adaptable adjacency matrices by combining candidate matrices in relation to different inputs.

7.1.3 Deep Learning Methodologies

As we discussed, many traditional methods are used for human motion prediction, getting back to trend using all data-driven deep learning models (DL), like convolution neural network (CNN), recurrent neural network (RNN), long short-term memory (LSTM), generative adversarial network

(GAN), and graph convolution neural network (GCN). Discussing CNN, [1–3, 6, 9, 13, 17], used end-to-end feed-forward network and trajectory convolution neural network to forecast future human poses. In this model, they used trajectory space and focus to model motion kinematics of input sequence using coupled spatiotemporal features, effective global–local features, and, last, global co-occurrences on temporal features in new space. Particularly, coupled spatiotemporal features describes the temporal and spatial structural data concealed in a natural pattern of motion by humans, which can be effortlessly extracted by the DL model CNN by concurrently covering the temporal and spatial features of all the sequences with the use of conventional filters. This model addresses an important aspect of predicting human motion from an image or video because human motion not only focuses on postures and joint movements of human parts but also on the entire local and global features in the data to get better prediction results, which are obtained through CNN. The work of Rajasekaran *et al.* on convolution neural network treats pose sequence as a matrix of two dimensions where each dimension indicates spatial axis and temporal axis, respectively. Collectively, the spatiotemporal convolution filters are used on images to pose data. The pose data are different from images, which are the essence of prediction and lack recurrent elements, which give more counter to the same filter; hence, this affects the accuracy and effectiveness of the convolution neural network model [18].

By working on the limitations in previous models, the recurrent neural network is an efficient deep learning model used for prediction as it has the advantage of dealing with tasks related to time series. Recurrent neural network models process consecutive data using a repeated link joining its neural activation at successive times. [14] apply recurrent network RNN for optical prediction by quantizing an optical signal into a glossary of optical or visual words and predicting a diffusion over those words in the next frame, given the optical sequence of a word observed at a specific pixel location. [15] particularly focused on human motion prediction, where all the likelihood future three-dimensional poses of a person providing past motions were forecasted. In Figure 7.1 below, using RNN models for prediction from past human poses, the model predicts the future pose of human motion. A recurrent neural network model has many more aspects working on the needs to attain certain objectives like the following:

- ➢ Deep recurrent neural networks (DRNN),
- ➢ Structural recurrent neural networks (SRNN),

- ➢ Encoder recurrent decoder (ERD),
- ➢ Generative recurrent unit (GRU),
- ➢ Long short-term memory (LSTM),
- ➢ Gated recurrent unit (GRU), etc.

There are various NN models addressing certain defects like working on long-term predictions, addressing spatial defects affecting prediction, and temporal strategy. Its explained in Figures 7.2, 7.3 and 7.4.

In [8] a generative adversarial networks produce realistic data with patterns that are similar to the training data. They still need expert training because they are fragile.

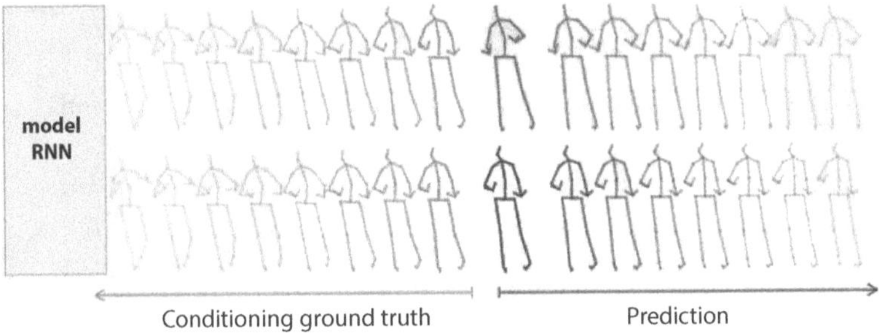

Figure 7.1 Future motion prediction using recurrent neural networks from past human poses in a sequence and obtaining future human motion.

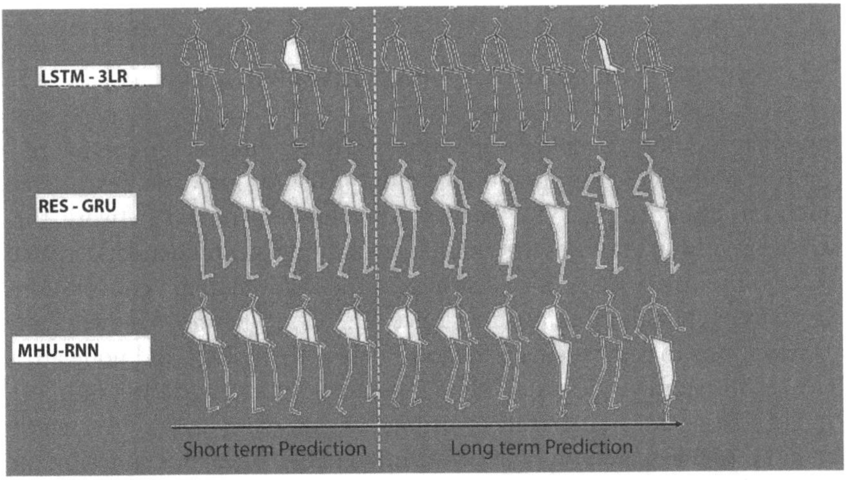

Figure 7.2 Various recurrent neural networks predicting human motion from short-term (SL) prediction to long-term prediction.

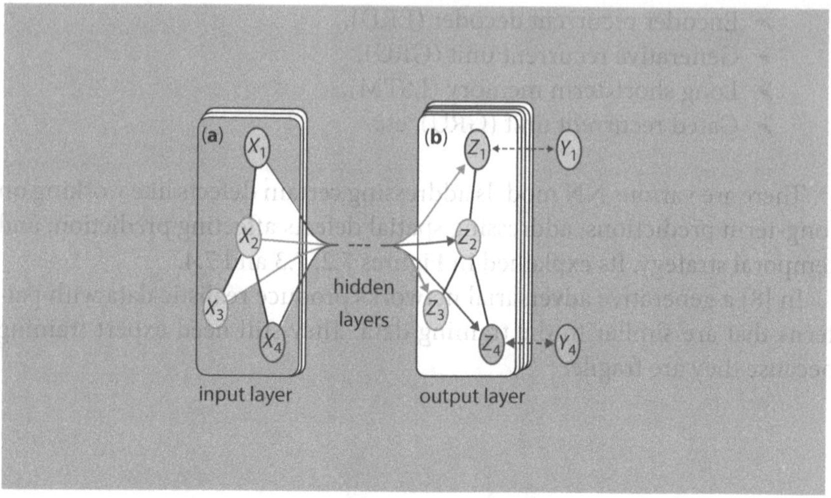

Figure 7.3 Simple representation of graph convolution (GCNN) neural network.

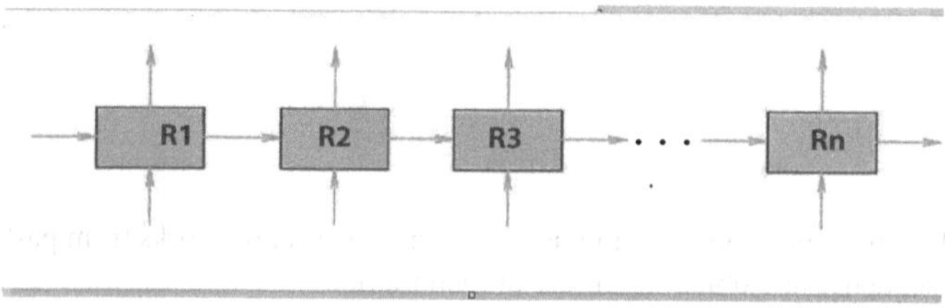

Figure 7.4 RNN on consecutive data.

7.2 Graph Convolution Neural Network (GCN)

An emerging deep learning semi-supervised learning technique for recognition and prediction, graph convolution neural networks learn from graph structured input. One sort of convolutional neural network that works directly on graphs is the graph convolutional neural network created by [16]. The native first-order nearness of the random graph convolutions has an impact on the architecture choice. The model develops hidden layer visualizations that encapsulate the local network topology and node attributes as the number of graph edges increases. According to [8] and [19], characteristics are transferred from fine to raw scales and subsequently from raw to fine scales. A graph convolution network can be trained on the relationships between any two human joints.

Graph convolution neural networks (GCNNs) simply rely on the data structure of a graph, which is more efficient in the application of dot languages or point analysis. We use this graph structure in human motion prediction because it takes the idea of using human skeletons joints as input, and using the spatial (spatio) and temporal movements of any human skeleton joint, the motion of any human can be predicted. This idea of taking the skeleton joint points is a very much efficient technique in human action recognition. Models used for humans doing recognition incorporate the skeleton point in the graph, which is applied to convolution filters to get the best feature extraction so that the accuracy rate of the recognition work model is greater than any other DL (deep learning) model in human action recognition. Taking advantage of using human skeleton joint points in motion prediction of humans is beneficial because the movement of one joint in the body can lead to some sort of pose or motion. The above-discussed models for several human motion predictions on the previous section, i.e., all variants of NN, etc., have many limitations on negation of kinematic or dynamic dependencies between human skeleton joints. The fully connected graphs constructed using links between every joint and its pair create a whole human motion predicting sequence.

7.2.1 Why Graph Convolution Network?

According to Li *et al.* [2], the most up-to-date approach presupposes that the human motion sequence as a whole comprises links between every joint pair in a completely connected graph. While encouraging performance has been achieved, superior accuracy results are generated since the graph convolution network ignores the underlying significant features of the natural connectivity of human skeleton joints. Conventional models, like convolutional neural networks, ignore the kinematic and dynamic dependencies of any human's skeleton joints. Graph convolutional neural networks (GCNNs) may be based on the concept of convolution neural networks (CNN), but GCNs are highly interpretable and offer improved generalizations over graphs. That is why GCNs take more attention and are widely applied in many fields. To crisply answer the question on why GCN and not other deep learning methodologies commonly used for prediction process, like convolutional neural network or recurrent neural network, which are also good for predicting, the simple answer is that convolutional neural networks train on grid-like structures, and recurrent neural networks train on consecutive data. In contrast, graph convolutional neural networks work on graphs, which are a highly unstructured data type.

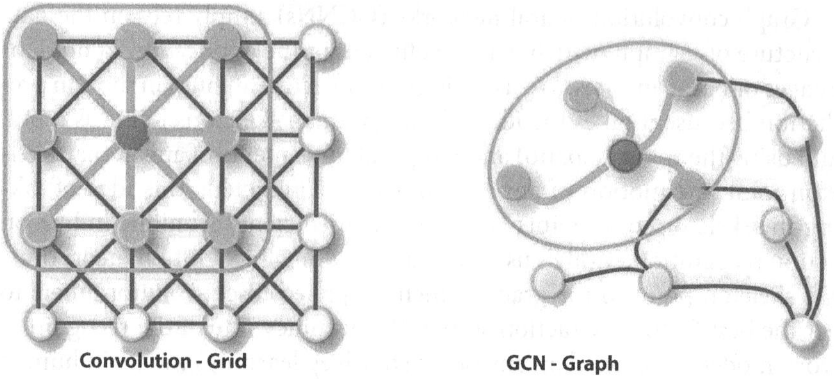

Figure 7.5 CNN train on the grid-like structure and GCN on graph structure.

Figure 7.4 shows how recurrent neural network trains and works on continuous or sequential data, and Figure 7.5 below provides you with better clarity in understanding convolutional neural networks (CNN) on a grid structure and graph convolutional neural networks (GCN) on a graph structure. Visualizing the structure will give a better understanding.

7.3 Forms of GCN on Human Motion Prediction

Graph convolutional networks being executed alone for human motion prediction leaves out many problems unanswered like not addressing spatial aspects, temporal aspects, and context statements; hence, there are many types of research that have contributed to fixing these problems concerning graph convolutional networks. In this section, we will be discussing various versions or forms of GCNs that contributed to human motion prediction.

7.3.1 Gated Graph Convolutional Network

A gated graph convolutional network is a super-lattice data passing-based graph network that uses edge gates, residual interconnections, and batch normalization. A gated graph convolution network allows input graphs with various structures so that the GGCN model will learn automatically without considering the graph architecture. Also, the GGCN model uses features for calculation similarity and structural properties of nodes in the graph where there is no need for a deep understanding of graphical architecture. Figure 7.6 depicts the architecture of GGCN.

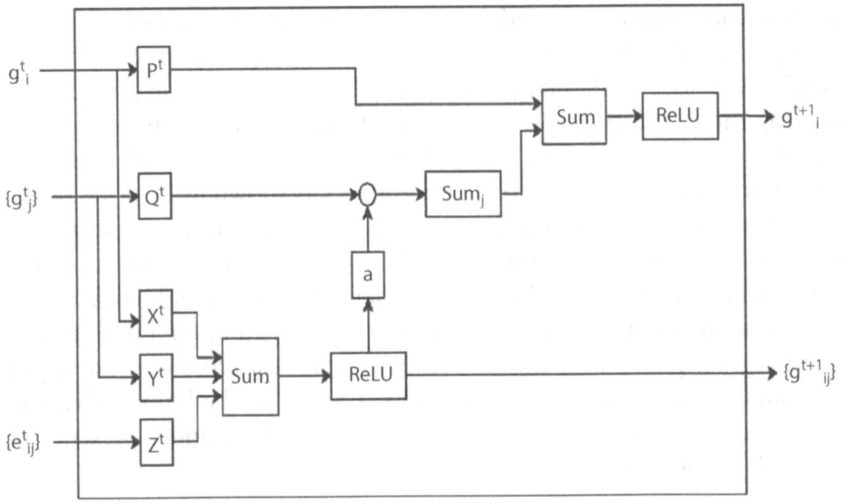

Figure 7.6 Architecture of a GGCN model.

$$g_i^{t+1} = g_i^t + \left(\left(P^t g^t + \sum{}^{i \in N} e_i^t \odot Q_i^t g_{ij}^t\right)\right)_j \tag{7.1}$$

Here, $P^t, Q^t \in R^{d \times d}, \odot$ implies element-wise multiplication, *Ni*, are neighbors of node *i*, and e^t $i_i s_j$ is the edge gate defined as follows:

$$= \left(e_{ij}^{t^{ij}}\right) \div \left(\sum j \in N\left(e_i^{t^{ij}}\right)\right) + \varepsilon \tag{7.2}$$

$$e^{ij} = e_{ij}^{t-1} + ReLU\left(BN\left(X^t g_i^{t-1} + Y^t g_j^{t-1} + Z^t g_{ij}^{t-1}\right)\right) \tag{7.3}$$

Here, σ t sigmoid function; also, ε is a small fixed constant for numerical stability, and $X^t, Y^t, Z^t \in R^{d \times d}$

7.3.2 Spatial and Temporal Graph Convolutional Network

Human motion prediction not only depends on the past motion sequence but also on monitoring the relations between various joints in the human skeleton. As we all know, the movement of joints defines the temporal attributes, and processing it based on computer vision, i.e., video-to-image frames, concentrates on spatial attributes. Due to the decoupled design strategy's inability to accurately express the cross-dependency of such temporal as well as spatial linkages, there may be a generalization problem. To develop the forecasting model to learn difficult spatial and temporal correlations over multiple data kinds, [9] suggested spatio (spatial)-temporal gating-adjacency graph convolution networks as shown in Figure 7.7.

Through the use of an adjacency matrix of adaptable trainable adjacency metrics that combines input spatial and temporal adjacency metrics, gating networks are utilized to enhance the generalization of graph convolution networks. By modifying the vast number of spatial and temporal models and integrating various spatial and temporal data, the cross-dependency of time and space is addressed. This model is based on the human skeleton, which we term cross dependency in this model. In this model, present two methods to forecast human motion.

- ➢ Enhancing strategy
- ➢ Balancing strategy

Zhong, C. *et al.* [9] used TCN as a decode tool after using GAGCN as an encoder to determine the spatiotemporal importance of the chronological motion sequence. To acquire the integrating coefficients {"w I s"} and {"w I t,"}, input the feature from the previous layer into two distinct

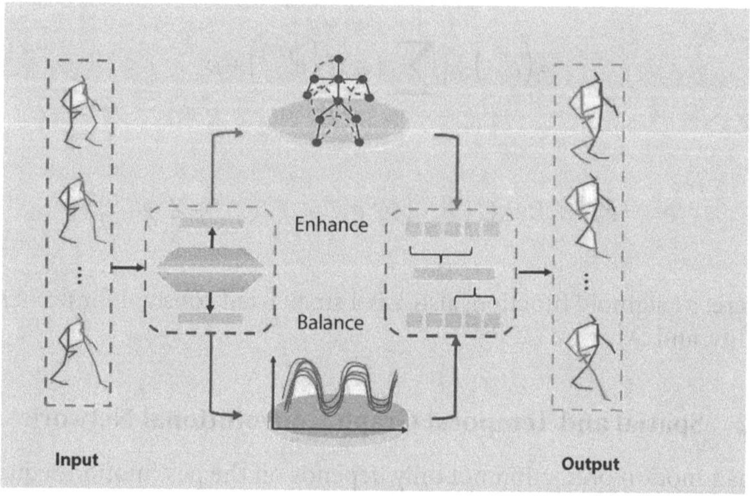

Figure 7.7 An overview of strategies employed in the spatiotemporal gating adjacency GCN.

gating networks in this order: a spatially gating network and a temporally gating network. The adaptive spatial (temporal) adjacency matrix is then created by mixing the spatial (temporal) adjacency matrix with the blending parameters. The spatial and temporal correlations are then combined with the (KP) Kronecker product to provide the characteristics for the subsequent layer. Movement is foreseen by encoder–decoder structures. Three components make up the Gating-Adjacency GCN (GAGCN) encoder model developed.

- To obtain the integrating coefficients which are the features from the chronological layers, the features are loaded into a spatially gating networks and a temporally gating networks, in the first part.
- Second, using the calculated integrating coefficients, we mix the spatial adjacency matrix with the temporal adjacency matrix to frame the adaptive adjacency matrix.
- Last, to produce all the features for the following layer, we finally combine the spatial and temporal dependencies with the (KP) Kronecker product having the latent motion for the decoder.

We employ the temporal convolutional networks (TCNs) to forecast the upcoming sequence after transferring through six GAGCN layers.

The representation of the pose is a graph G = (V, E) based on the skeleton, where E is the edge set and V is the joint set. The nodes of joint features are three-dimensional coordinates or joints, and the edges have a relationship with the adjacency matrix $K \in V^{Y \cdot Y}$.

$$I^{l+1} = (I^l; A; W) = (KI^l W^l) \tag{7.4}$$

where, $K, I^l \in V^Y$ and $W^l \in V^{F_l \cdot F_{l+1}}$

- Spatiotemporal gating

$$\{\omega^i\} = (I) = softmax(FC(I)) \tag{7.5}$$

Here, FC implies the three fully connected (FC) layers, *softmax* is the (AF) activation function, I is the input feature, and $\{\omega^i\}$ is a set of integrating coefficients.

$$K = \sum_i K^i.\omega^i \tag{7.6}$$

where $\{K^i\}$ is a set of trainable adjacency matrices and K is the adaptive adjacency matrix.

- Spatiotemporal (STM) modeling

Here, division of the adjacency matrix K into Ks and Kt occurs.

$\in V^{Y \cdot Y}$ depicts the connections between joints, whether they are connected or not by a skeletal link. After learning the integrating coefficients for the spatial and temporal features based on Equations 7.2 and 7.3, two gating networks blend the candidate adjacency matrices to create the adaptive adjacency matrix. These two equations are updated further to take the following shape:

$$\left\{\omega_k^i = Gating_k^{(I)}, A_k = \sum_i^q K_k^i.\right\}_k \tag{7.7}$$

7.3.3 Dynamic Multi-Scale Spatiotemporal Graph Convolution Network

Another type of graph convolutional network covering spatial and temporal variables on the domain of human motion predictions will be covered in this section. This prediction model is constrained by the shortcomings of earlier work on recurrent neural network models, such as mistake accumulation that results in unreliable results. Since multiple scales fuse dynamically during training, a multi-scale spatiotemporal graph convolutional network (GCN) was proposed by [17] to imply the spatial a temporal dependencies in the process on human movement. The entire concept adheres to an encoder–decoder structure and is appropriate for all actions. The encoder employs semi-autonomous learning spatial GCNs and temporal GCNs that record motion characteristics between frames to extract spatial structure from joint trajectories. Temporal convolution networks (TCNs) are used by the decoder to preserve its robust capability.

In this, the technique of a mask-restricted semi-autonomous self-learning graph convolution is produced. To prevent information loss, using this approach, the human dynamic chain is prevented from becoming contaminated by joint inputs. Additionally, it blocks all the networks from

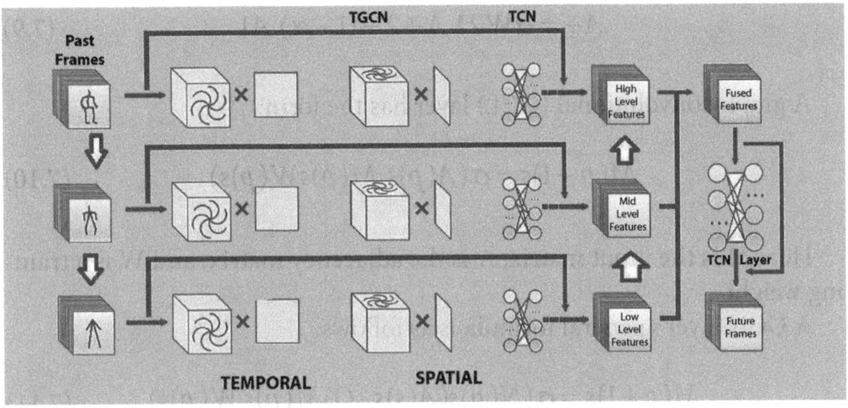

Figure 7.8 Framework of a dynamic multi-scale spatiotemporal GCN.

convergence to the (local) non-global optimum by closing the data transfer channels of remote joints, and it teaches networks the link quality among joints automatically. Since recurrent neural networks are used to lose information during transmission, the proposed model's temporal convolution network achieved the best results on many tasks because of parallel training by a flexible receptive field.

Figure 7.8 provides a visual architecture of an (dynamic) active multi-scale spatiotemporal graph convolution network. The model must forecast the stance for F upcoming frames based on the pose of a body determined by the three-dimensional coordinates of T number of human joints for N frames. The observed pose sample can be written as A1: N = a1, a2, ... an, ..., aF, where a frame is an x at a time and 1 = n = N. The prediction frames are thus designated as Bn + 1: n + F = Bn + 1, Bn + 2, Bn + F, while the matching ground truth is designated as Bn + 1: b + n = Bn + 1, Bn + 2, Bn + F. X, C, N, and T-shaped tensor are the input. Graph Gs = (Vs, Es) in spatial graph convolution, using V nodes I Vs. A concealed spatial adjacency is written as R V × V for edges I j of type Es. To build a different graph with T nodes t Vt for temporal graph convolution, we define Gt = (Vt, Et). A temporal adjacency At R V × V is used to express the temporal edges (t, u) Et.

Dynamic Fusing: The following formula can be used to describe a fusing process:

$$A+2=\alpha W32\ A3+(1-\alpha)\ A2 \tag{7.8}$$

$$A+ = \alpha W21\,A + 2 + (1-\alpha)\,A1 \tag{7.9}$$

A graph convolutional (GCL) layer has the form:

$$M(p+1)s = \sigma(A(p)s\ M(p)sW(p)s) \tag{7.10}$$

Here M is the input matrix, A is the adjacency matrix, and W is a training weight.

A GCN layer's general formula is as follows:

$$M(p+1)s = \sigma(N(p)s(A(s)s \circ O)M(p)s\ W(p)s) \tag{7.11}$$

Also, here, O is a concealed matrix to block the inter-connection of any joint pairs that are far in distance.

- Temporal GCN: adds any spatial-sharing weight table only, N (p) t, containing the temporal features that are learned:

$$M(p+1)t = \sigma(N(p)t\ A(p)\ t\ M(p)\ t\ \ W(p)\ t) \tag{7.12}$$

Here, $M\ (p)\ t \in R\ N \times C\ (p)$, which are the input matrices of layers p, and $W(p)\ t \in R\ C\ (p) \times C\ (p+1)$ are the weight trainable matrices of layer p.

7.3.4 Multi-Scale Residual Graph Convolution Network (MSR-GCN)

Different graph convolutional networks have demonstrated to be particularly effective in learning the dynamic and kinetic relationships among the joints of all human stances, which are very helpful in predicting human motion. Moreover, the abstract of a human posture frequently produces a collection of motion poses at various scales. As the level of abstraction is raised, the pose's motion becomes stable, which helps motion prediction as well. [8] suggested a brand-new multi-scale residual graph convolution network (MSR-GCN) in forecasting human motion. This MSR-GCN is used to model the position as a fully connected graph of the human skeleton with motion joints as the nodes. GCN is then used to learn associations among all pairs of skeleton joints, irrespective of how close together they are in reality. However, the whole structure of the human stance cannot be captured by the graph convolution network alone. A human stance can be abstracted, by grouping neighboring joints and together expressing the cluster by only one joint producing a harsher pose. Since several joints usually originate from one body part, gradually abstracting different body parts in this way can significantly stabilize the body's motion. We discover that the motion is more stable at the coarser level making position prediction simpler. It is encouraging to begin predicting poses at the most basic level and work your way up to the finer ones afterward.

[8] explained in Figure 7.9 about MSR-GCN which comprises of a start initial GCN,

- Descending GCNs (De0, De1, De2, De3),
- Ascending GCNs (As0, As1, As2, As3), and
- End GCNs (Ed0, Ed1, Ed2, Ed3).

A black input is used in the first start graph convolution network. The next step is to pile simultaneously De0, De1, De2, De3, and As0, As1, As2, and As3 GCNs for feature extraction on each scale twice. The final step is to feed the corresponding end GCNs with the combination of features at each scale for decoding. The remaining connections are built on each final end GCN that adds the fundamental poses to each GCN's output teaching the network underused poses instead of the desired poses directly.

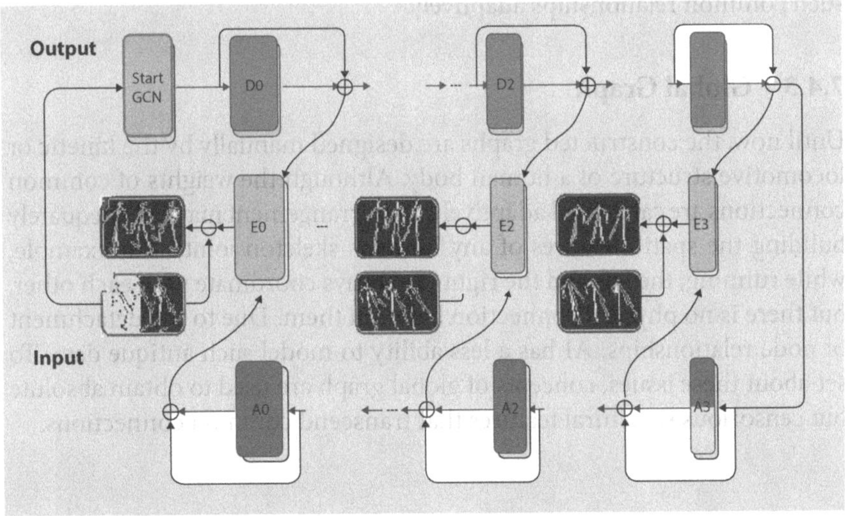

Figure 7.9 Architecture of MSR-GCN.

7.4 Types of Graphs Employed on GCN

For human motion prediction, various graph structures are employed to adapt the human skeleton joints to predict. In this section, we will discuss a few graphs induced in the human motion prediction process.

7.4.1 Skeleton-Based Pose as an Undirected Graph

This graph has the deepening of the network's grading with persistent depiction and may also partly style high-level features. Yet, this policy is not a perfect expression. Viscerally, all the skeletal joints that are connected with the vertices contribute unequally to the motion or movement pattern for the vertex. For example, during motion, like walking, the shoulder joint in the human skeleton is more of a hanger on the trunk than on the elbow joint.

7.4.2 Connective Graph

To resolve the problem from the previous example, the connective graph constrained by the adjacency matrices provides the representation of the learnable connections' robustness of the common connection on a human skeleton. Despite using the constant weight of the adjacency matrix, the interdependencies of common connections on human skeletons that are trained, which is better than using the constant, finally learn the weights of such common relationships adaptively.

7.4.3 Global Graph

Until now, the constructed graphs are designed manually by the kinetic or locomotive structure of a human body. Although the weights of common connections are calculated adaptively, this arrangement may fall adequately building the spatio features of any human's skeleton joints. For example, while running, the left and the right leg always coordinate with each other, but there is no physical connection between them. Due to the detachment of node relationships, AI has a less ability to model such antique data. To set about these issues, concepts of global graph are used to obtain absolute but censorious structural features that transcend common connections.

7.5 Conclusion

By summarizing the concept of graph convolution network, predicting human motion as our goal, we skimmed a few concepts of other deep learning techniques so far used in this research domain and had a brief discussion on graph convolution network and its various forms addressing defects that bother the results while using graph convolution networks (GCNs). Graph convolutional network models have produced better results on human motion prediction as it takes the structure of graph data structure, which incorporates human skeleton points as nodes and edges in the graph, unlike grid-like structures, which are used by other deep learning techniques. Prediction processes on human motion are a quite complex process than human action recognition; hence, depending upon the statement of a problem formulated or issues that need to be resolved, the usage of deep learning models is chosen. On the research made on the topic of graph convolution network, this model is highly suitable for inputs with point or dot dependencies; hence, GCNs can be incorporated into human motion prediction following human skeleton joint points. Further in the future, various versions of GCNs may be introduced for motion prediction providing progress in this field.

References

1. Malini, A., Priyadharshini, P., Sabeena, S., An automatic assessment of road condition from aerial imagery using modified VGG architecture in faster-RCNN framework. *J. Intell. Fuzzy Syst.*, 40, 6, 11411–11422, 2021.
2. Li, M., Chen, S., Zhao, Y., Zhang, Y., Wang, Y., Tian, Q., Dynamic multiscale graph neural networks for 3d skeleton-based human motion prediction, in: *Proceedings of the IEEE/CVF Conference on Computer Vision and Pattern Recognition*, pp. 214–223, 2020.
3. Liu, Z., Zhang, H., Chen, Z., Wang, Z., Ouyang, W., Disentangling and unifying graph convolutions for skeleton-based action recognition, in: *Proceedings of the IEEE/CVF Conference on Computer Vision and Pattern Recognition*, pp. 143–152, 2020.
4. Mao, W., Liu, M., Salzmann, M., Li, H., Learning trajectory dependencies for human motion prediction, in: *Proceedings of the IEEE/CVF International Conference on Computer Vision*, pp. 9489–9497, 2019.
5. Li, F., Zhu, A., Xu, Y., Cui, R., Hua, G., Multi-stream and enhanced spatial-temporal graph convolution network for skeleton-based action recognition. *IEEE Access*, 8, 97757–97770, 2020.

6. Shi, L., Zhang, Y., Cheng, J., Lu, H., Skeleton-based action recognition with directed graph neural networks, in: *Proceedings of the IEEE/CVF Conference on Computer Vision and Pattern Recognition*, pp. 7912–7921, 2019.
7. Zhang, X., Xu, C., Tao, D., Context-aware graph convolution for skeleton-based action recognition, in: *Proceedings of the IEEE/CVF Conference on Computer Vision and Pattern Recognition*, pp. 14333–14342, 2020.
8. Dang, L., Nie, Y., Long, C., Zhang, Q., Li, G., MSR-GCN: Multi-Scale Residual Graph Convolution Networks for Human Motion Prediction, in: *Proceedings of the IEEE/CVF International Conference on Computer Vision*, pp. 11467–11476, 2021.
9. Zhong, C., Hu, L., Zhang, Z., Ye, Y., Xia, S., Spatio-Temporal Gating-Adjacency GCN for Human Motion Prediction, in: *Proceedings of the IEEE/CVF Conference on Computer Vision and Pattern Recognition*, pp. 6447–6456, 2022.
10. Cui, Q., Sun, H., Yang, F., Learning dynamic relationships for 3d human motion prediction, in: *Proceedings of the IEEE/CVF Conference on Computer Vision and Pattern Recognition*, pp. 6519–6527, 2020.
11. Jordan, M., II and Jacobs, R.A., Hierarchical mixtures of experts and the EM algorithm. *Neural Comput.*, 6, 2, 181–214, 1994.
12. Taylor, G.W., Hinton, G.E., Roweis, S., Modeling human motion using binary latent variables, in: *Advances in Neural Information Processing Systems*, vol. 19, 2006.
13. Liu, X., Yin, J., Liu, J., Ding, P., Liu, J., Liu, H., Trajectorycnn: a new spatiotemporal feature learning network for human motion prediction. *IEEE Trans. Circuits Syst. Video Technol.*, 31, 6, 2133–2146, 2020.
14. Ranzato, M., Szlam, A., Bruna, J., Mathieu, M., Collobert, R., Chopra, S., Video (language) modeling: A baseline for generative models of natural videos. *Cornell University Journal*, 2014. *arXiv preprint arXiv:1412.6604.*
15. Martinez, J., Black, M.J., Romero, J., On human motion prediction using recurrent neural networks, in: *Proceedings of the IEEE Conference on Computer Vision and Pattern Recognition*, pp. 2891–2900, 2017.
16. Kipf, T.N. and Welling, M., *Semi-supervised classification with graph convolutional networks*, 2016, *arXiv preprint arXiv:1609.02907.*
17. Yan, Z., Zhai, D.H., Xia, Y., *DMS-GCN: Dynamic Multiscale Spatiotemporal Graph Convolutional Networks for Human Motion Prediction*, 2021, *arXiv preprint arXiv:2112.10365.*
18. Rajasekaran, U., Sriram, G.K., Ramyavarshini, P., Malini, A., XAI-Based Light-Weight CNN-HAR Model Using Random Sampling, in: *International Conference on Emerging Trends and Technologies on Intelligent Systems*, Springer Nature Singapore, Singapore, pp. 377–388, February 2023.
19. Subbulakshmi, B., Nirmaladevi, M., Rajeswari, A.M., Sivakumar, M., A Graph-Based Approach For Personalized Job Recommendation. *BioGecko*, 12, 3, 4474–4486, May-June 2023. http://biogecko.co.nz/.2023.v12.i03.

20. Nirmala Devi, M., Harsha Prada, M.D., Sri Saila, A.S., Subbulakshmi, B., Semi Supervised Approach with Entity Embeddings for Heart Disease Prediction, in: *Lecture Notes in Networks and Systems*, vol. 671, Springer, Cham, Scopus Q4, May 2023, https://doi.org/1007/978-3-031-31153-6_44.
21. Subbulakshmi, B., Nirmala Devi, M., Sriram, Srimadhi, Arvindhan, M., A Hybrid Machine Learning Model for House Price Prediction. *Intelligent Manufacturing Systems in Industry 4.0. IPDIMS 2022. Lecture Notes in Mechanical Engineering*, Springer, Singapore, Scopus Q4, Jul 2023, https://doi.org/10.1007/978-981-99-1665-8_35.

8

GraphChXNet: A Graph Convolutional Neural Network-Based Model for Detecting Chest Diseases Using X-Ray Images

D. Kiruthika*, N. Vinothini, G. Jegan and G. Ananthi

Thiagarajar College of Engineering, Madurai, India

Abstract

Chest diseases pose a significant threat to global health, and timely diagnosis is crucial for effective treatment. In recent years, advancements in convolutional neural networks (CNNs) have enabled machine learning algorithms to assist healthcare professionals in identifying various medical conditions through analysis of chest X-ray images. These advanced systems rely on the unique features of tissue textures and shapes visible in diagnostic imagery to make accurate projections about patient health. Recent research has leveraged large collections of high-quality chest X-rays along with cutting-edge deep learning techniques to enhance the accuracy of disease diagnosis in this field. Indeed, various investigations have sought to enhance existing deep learning techniques by incorporating multiple scales of imaging characteristics from diagnosis chest X-rays. Unfortunately, these efforts often neglect the computational burden and sacrifice spatial accuracy to cover a broader field of view when trying to extract meaningful features from high-resolution chest X-ray images. In this study, we explore a novel approach in predicting chest diseases by treating it as a graph node classification problem. Specifically, we represent individuals within a potential disease population as nodes on a graph, with edges denoting relationships between them. Using a graph convolutional neural network (GraphChXNet) model, we then extract graph embeddings that capture both the topological structure of the graph and the characteristics of individual nodes. These embeddings are further refined *via* an attention mechanism that learns to focus on the most important weights among those captured by the model. To handle varying scenarios, we also incorporate a K-nearest neighbor graph method to model the connection between the disease population and

**Corresponding author*: dkiruthika@student.tce.edu

Malini Alagarsamy, Rajesh Kumar Dhanaraj, J. Felicia Lilian, Vandana Sharma and Gheorghita Ghinea (eds.) Graph Convolutional Neural Networks for Computer Vision, (161–180)

its surrounding neighbors. In this study, we assessed the effectiveness of various components within a machine learning model in predicting chest diseases using an experimental approach. We evaluated how these components influenced the model's performance and compared it to benchmarks. The novel approach outperformed other models on the NIH Chest disease dataset, enhancing accuracy and sensitivity. The CNN model achieved 94.2% accuracy and 5.7% recall, with a slight downgrade in F1 score compared to non-graph targeting models.

Keywords: Chest diseases, graph neural network, KNN methods, x-ray images

8.1 Introduction

Medical imaging technology, especially chest X-ray analysis, has advanced dramatically in recent years, transforming the diagnosis and management of respiratory conditions. With the help of sophisticated algorithms and machine learning, healthcare providers now rely on chest X-ray pneumonia diagnosis. Graph-based models like CGNet have also been proposed to embed prior knowledge and improve detection accuracy for pneumonia using X-rays [1]. Integrating cutting-edge tools, such as deep learning and artificial intelligence, that can detect a wide range of respiratory diseases, such as tuberculosis and pneumonia in cancer, has made chest X-ray examination more accurate and effective and easier to perform in this area. One notable development is CheXNet, an AI system that automates a radiologist-like function to diagnose lung disease in chest X-ray. These remarkable advances demonstrate the tremendous potential of combining machine learning and medical imaging to improve patient care and outcomes [2]. Using state-of-the-art machine learning technologies, CheXNet has demonstrated an impressive ability to match the diagnostic skills of pneumonia detected on chest X-ray. This improvement is a wonder that greatly relieves the overburdened healthcare systems and the incredibly convenient and accurate method for early diagnosis. By leveraging sophisticated convolutional neural networks (CNNs) and extensive storage of chest X-ray images, CheXNet has raised the standard of radiologist-level performance in lung disease diagnosis [3].

The emergence of attention-based graph neural networks (AGNNs) represents a major advance in semi-supervised learning, especially when dealing with large complex data sets. In principle, AGNNs use various forms of focus to optimize graph-structured data making, where nodes and edges cover microscopic relationships. When faced with limited labeled data, and the goal is to measure information of unlabeled observations, AGNNs exhibit exceptional performance. With the ability to selectively focus on relevant nodes in a graph, AGNNs efficiently extend knowledge, improving prediction accuracy [4]. Class-based focus in medicine in image analysis represents an important advance in the application of artificial intelligence (AI) in healthcare. This new approach addresses the challenge of classifying multiple lung states from chest images with unprecedented accuracy and efficiency. Using a class-specific conceptual model, the system enables the diagnosis and classification of various diseases, such as pneumonia, tuberculosis and lung cancer, which are often manifested by various symptoms within a single image Dyogi's influence is great because it enables a detailed assessment of the condition of the patients, and it was perfect [5]. The ability to recognize subtle symptoms of disease enables physicians to deliver better personalized treatment plans for individual patients, improving patient outcomes and optimizing healthcare with increasing use of AI in the field of medicine.

The integration of graph neural networks and attention mechanisms into recommendation systems enables the analysis of existing user interactions and the transfer of preferences in personal sessions This new approach is particularly useful in tasks such as in e-commerce, content streaming, and online services, where user interest is accelerated by targeting content or the most appropriate content using focus. Thus, these systems improve the accuracy of recommendations. Combined with the growing need for customized digital platforms and state-of-the-art machine learning, the recommended presentation of personalized graph neural networks with focused attention in conferences represents a significant advance in modern recommendation systems [6].

In the field of artificial intelligence, deep learning with graph convolutional networks (GCNs) is a rapidly growing field that shows great promise as GCNs have proven to be highly scalable and usable effectively across multiple disciplines from node classification to link prediction and beyond. Pushing the boundaries of manufacturing has played an important role. Moving forward, these networks are likely to transform the business once more by enabling complex and nuanced data analytics as artificial intelligence evolves, particularly with the adoption of deep approaches like graph convolutional networks (GCNs), which are an increasingly integrated digital ecosystem that provides novel solutions [7]. The ongoing struggle toward the COVID-19 pandemic has caused the improvement of sophisticated technologies aimed toward improving prognosis. One such groundbreaking software is the use of GCNs for detecting COVID-19 from chest X-ray images and patient statistics representing an extensive leap forward in medical imaging evaluation [8]. Using graph convolutional networks (GCNs) to investigate each X-ray pix and vital patient information, together with clinical histories and demographics, offers an intensive and statistic-pushed method for identifying COVID-19. By combining these records, GCNs can more accurately identify diseases, enhance patient treatment, and assist in public health initiatives at some stage in high-quality conditions brought about by the pandemic. This demonstrates how interdisciplinary techniques may be utilized in medicinal drug to increase diagnostic accuracy, streamline affected person care, and in the end benefit people's standard well-being [9].

The contribution of the paper are as follows:

- A novel GraphChXNet version is proposed comprising a graph convolutional neural network (GCNN) technique, which has shown outstanding performance in classifying nodes, to address the complex undertaking of chest disorder diagnosis. By leveraging the strengths of GCNNs, this study aims to improve upon current techniques and offer more accurate predictions concerning the uncertainty stages of chest diseases.
- The proposed version is in comparison with the conventional deep learning algorithms, like Naive Bayes, random forest, and support vector machine, to detect chest illnesses.
- In this paper, we as compared and contrasted two tactics in creating graphs that represent the relationships among people with chest diseases. Rather than relying on a single graph structure, those methods allow us to create personalized similarity networks for each patient primarily based on their specific traits and clinical history.

8.2 Proposed Methodology

In recent times, graph convolutional neural networks (GCNs) have shown amazing promise in addressing essential medical demanding situations, particularly in picture and non-photo scenarios. The core of these GNN fashions includes updating the representations of nodes with the aid of aggregating information from their neighboring nodes through repeated iterations. As undirected graph tasks frequently involve both supervised and semi-supervised mastering, GCNs are well suited for tackling such problems. Any unstructured graph can be described as a set of sections, G = (A X), with A as an adjacency matrix that represents the connections between nodes in the graph, and every access A_{ij} has a value of 1 if there is an edge between nodes and j, and zero otherwise. The matrix contains dimensions N × N, where N represents the number of nodes in the graph. X is a vector all functions related to each node in a graph, wherein Xi has dimensions N × D, and D represents the range of capabilities for each node. In summary, the representation of the graph as G = (A X) includes both the graph's structural data (represented by the adjacency matrix A) plus the nodes' attribute statistics.

$$HI^{(i+1)} = ReLU\left(\tilde{D} * \tilde{A}\tilde{D} + HI^{(i)}W^{(l)}\right) \quad (8.1)$$

A convolutional layer generates an output sequence of size $H^{(+1)}$, where H(0) = X initially. In the context of an undirected graph, the adjacency matrix A and identity matrix I are equivalent to each other and are both identical to A. The diagonal matrix D incorporates simplest non-zero factors on the principal diagonals. To perform a supervised node classification task using a layered graph convolutional network (GCN) structure, we can represent the very last embedded representation of the input nodes as follows:

$$Z = f(Xi, A) = softmaxReLU\left(\left(\hat{A}\, XiW^{(0)}\right)W^{(1)}\right) \quad (8.2)$$

In Equation 8.2 of the supplied textual content, it is mentioned how a graph convolutional network (GCN) model works. Specifically, it is defined that the GCN version takes as input a fixed d-dimensional features representing an item's attributes that are fed into a hidden layer with weights described by the matrix W(1) ∈ R nhid × C. This hidden layer is then used to compute the probability distribution over the C instructions present within the information, where each class is represented by way of a vector Xi. To do this, the authors use the softmax characteristic, which normalizes the values in the vector Xi to ensure they sum to 1 for every class. After training the GCN model on the input features and corresponding label information, the resulting outputs or predictions are obtained in the form of labels Y.

A. Population Graph Construction

Using phenomenological knowledge, researchers depict entire populations as sparse network topologies represented by graphs. To identify the most informative characteristics in the training dataset, a classifier is employed because of the high dimensionality of the connectivity matrix. By analyzing data from the National Institutes of Health (NIH), we generated a population graph shown in Figure 8.1. The population graph G, consisting of N nodes representing the entire population of patients, is constructed using edges E connecting these nodes. Each edge has a weight W associated with it. In total, there are N × N edges in the graph, where N is the number of nodes or patients. Additionally, there are P phenotypic important measures, represented by the set E = $\{P_h\}$, which serve as the edges of the graph. Finally, the adjacency matrix A_p of the graph is defined based on the weights of these edges.

$$A(x_i, x_j) = Exp\left(-\frac{[\rho(x_i, x_j)]^2}{2\sigma^2}\right) \cdot \sum_{h=1}^{H} E(P_h(i), P_h(j)) \tag{8.3}$$

Based on the exponential function, the distance between graph nodes is calculated and used to determine the weight of edges between them. When dealing with an incomplete graph, the edge weight between nodes that are most similar to each other is increased. The similarity between phenotypic measures is measured using a correlation coefficient, which determines the width of the kernel. This means that affected individuals grouped into the same class (both low or excessive threat) seem to have greater comparable node structures than those in exclusive categories.

$$E(P_h(i), P_h(j)) = \{1, if \mid \mu_i - \mu_j \mid < \beta\, 0, otherwise. \tag{8.4}$$

The weight matrix $E(P_h(i), P_h(j))$ in Equation 8.4 encodes the proximity or dissimilarity among pairs of phenotypically tremendous capabilities or indicators within the graph. The size of this matrix relies upon the unique feature being analyzed, with larger matrices representing more complicated relationships among functions. Additionally, the weights within the matrix are determined primarily based on the metric used to outline importance, with qualitative measures resulting in a matrix wherein all elements are either 0 or 1, while quantitative measures use a unit step characteristic to assign weights primarily based on the gap among elements. This permits edges to be assigned more weight while connecting nodes with comparable phenotypic profiles and reduces the load of edges connecting nodes with awesome profiles. By adjusting the cost of β, the power of those differences can be managed taking into account first-rate-tuning of the graph structure. Ultimately, evaluating and optimizing these similarity measures and phenotypic measurements *via* experimentation will enable the advent of extra informative and established graphs.

B. KNN Graph Construction

In sure situations where graph facts lack organization or lack values, kernels close to neighbors (KNNs) are often applied in graph convolutional neural networks (GCNs). To create a preliminary graph earlier than extracting characteristic embeddings from GCN, a K-nearest neighbor community has been evolved. The technique of finding the most comparable neighbors within a specified distance threshold is known as KNN clustering. We establish connections between the top K similar neighbors for each example by linking them. If the likelihood of the probability density function parameters is uncertain, K-nearest neighbor categorization may be employed to carry out discrimination analysis. By evaluating the resemblance of node characteristics using cosine similarity approach, which primarily calculates the similarity between nodes' x_i and x_j features, we calculate the closeness between nodes.

$$S_{i,j} = \frac{X_i \cdot X_j}{|X_i||X_j|} \tag{8.5}$$

In step 5, we create a similarity matrix S ∈ R^N × N. Each element S(i,j) indicates the cosine similarity between the vector representation for node and the other nodes in the dataset omitting node itself. The cosine similarity measures the degree to which two vectors are oriented similarly in space, with higher values indicating a greater similarity. As the cosine value increases, the angle between the two vectors decreases, signifying a stronger connection. By recording these cosine similarities for each node, we may identify the top k most comparable node pairs and create edges connecting them, thereby forming an undirected K-nearest neighbor graph. Each node will have k adjacent nodes that share the most similar features. Finally, we represent the aggregated edge information as an adjacency matrix AF.

C. Proposed Model

The GraphChXNet model has limitations when it comes to integrating topological and node attributes. To address these challenges, the attention-based Multimodal GCNN was developed. This architecture employs three primary channels as follows:

1. Original Graph Channel: The GCNN can be trained using the entire population graph.

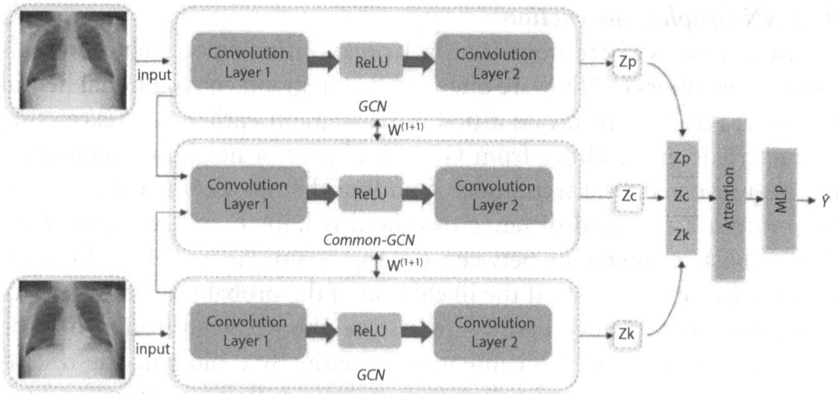

Figure 8.1 GraphChXNet framework.

2. Feature Graph Channel: The K-nearest neighbors graph is used to train the GCNN on this channel.
3. Information Graph Channel: Both the original graph and feature graph are utilized to train the GCNN on this channel. These three channels are combined to produce an improved representation of the graph for node classification tasks. The embeddings from each channel are concatenated with learned weight coefficients followed by a final embedding layer to make predictions. The GraphChXNet framework is shown in Figure 8.1.

i) Multi-Channel Graph Input

In this segment, it is proven that a graph construction can be applied to generate populations and K-nearest neighbor graphs, which function as the preliminary topological graph shape for the attention-primarily based message passing neural network. Specifically, they define the way to construct those graphs using the adjacency matrices received from the topological space and characteristic space, respectively. To create a population graph, the authors input the graph (Ap, X) into the version, wherein Ap is the adjacent matrix acquired from the topological area, and X represents the nodes' features. The embedded visualization of nodes on the (l+1)th level is specified as $Z_p^{(l+1)}$. In contrast, a K-nearest neighbor graph is generated by putting the graph (Af, X) in its modified version, where Af is the adjoining matrix retrieved from the feature area, while X reflects the nodes' features. The embedded form of the nodes within the (l+1)th layer is subsequently defined as $Z_f^{(l+1)}$. The authors observed that there is a relationship between the statistics in the topological area and feature space, which are not absolutely separate from one another. To bridge this gap, the authors added a shared channel with parameter sharing to the version, which they refer to as $Z_c^{(l+1)}$. The genuine calculation manner is described below.

$$Z_p^{(l+1)} = \left(\tilde{D}_p^{-\left(\frac{1}{2}\right)} \tilde{A}_p \tilde{D}_p^{-\left(\frac{1}{2}\right)} Z_p^{(l)} W_p^{(l+1)} \right) \tag{8.6}$$

$$Z_f^{(l+1)} = \left(\tilde{D}_f^{-\left(\frac{1}{2}\right)} \tilde{A}_f \tilde{D}_f^{-\left(\frac{1}{2}\right)} Z_f^{(l)} W_f^{(l+1)} \right) \tag{8.7}$$

$$Z_c^{(l+1)} = \left(\frac{Z_p^{(l+1)} + Z_f^{(l+1)}}{2} \right) \tag{8.8}$$

ii) Attention Mechanism

The three variables—Zp, Zf, and Zc—have been incorporated into the channel. To assess the significance of label assignations for nodes (ap, af, ac), which may relate to these variables individually or in combination, an attention mechanism has been implemented as follows:

$$(a_p, a_f, a_c) = Attention(Z_p, Z_f, Z_c) \tag{8.9}$$

The attention weights for each of the n nodes are represented by a_p, a_f, a_c, where these values lie within the set Rn × 1. Moreover, the value range of a is between 0 and 1. Each node has an embedding in the ith row of Z_p, which can be represented as $Z_{ip} \in R^{1\times h}$, where h′ is a positive integer. To derive the attention value ω_p^i, we apply a nonlinear transformation to the embedding using the function q, which maps the embedded vector to a scalar value, which is represented as,

$$\omega_p^i = q^T \cdot tan\left(W_p \cdot \left(Z_p^i \right)^T + b_p \right) \tag{8.10}$$

In Equation 8.10, the weight matrix $W_p \in R^{h' \times h}$ represents the learned weightings between the input features and the projected space, where h′ is the dimensionality of the projected space. Similarly, the bias vector $bp \in R^{h' \times 1}$ represents the learned biases for the projection. To obtain the attention weights for other layers, such as the embedding matrices Z_f and Z_c, we follow a similar approach by computing the attention value ω_i for each node in these matrices using a dot product attention mechanism. We next normalize those attention values with the softmax function to get the ultimate attention weight. These attention weights are utilized to compute the attention layer's final output, which is created by combining the embeddings and attention weights *via* element-wise multiplication, yielding the output vector Z_a. In addition, here is an attainable weight vector a ∈ R n × 1, wherein n is the number of nodes of the graph, and its elements denote the importance of each node. The weighted vectors is diagonalized to produce the last attention layer output.

$$Z_a = a_p \cdot Z_p + a_f \cdot Z_f + a_c \cdot Z_{c'} \tag{8.11}$$

iii) Objective Function

In this study, we employed a supervised binary classification task utilizing a linear transformation followed by a softmax function on the output of an attention mechanism, specifically the attention layer from Equation 8.11. The goal was to predict the classification label $\hat{Y}$ by determining the probability $\hat{Y}i$ that each node belongs to class C. To obtain this, we applied a linear transformation to the output of the attention layer, resulting in the converted output, which was then handed *via* the softmax function to acquire the probabilities Z_a. Finally, the anticipated type label was decided based on those possibilities.

$$\hat{Y} = softmax(W \cdot Z_a + b) \tag{8.12}$$

The AM-GCN model predicts an estimate or forecast of the true labels for every piece of data, which might be represented as Yl. In comparison, the actual true labels for those portions of data are denoted through L. To examine the accuracy of the model's predictions, we calculate the go-entropy errors between the anticipated and genuine labels across all training nodes. This degree presents us with a quantitative assessment of how well the model performs at classifying nodes in terms of their proper labels.

$$L_t = -\sum_{l \in Y_L} \sum_{i=1}^{c} Y_l ln \hat{Y}_l \tag{8.13}$$

8.3 Results and Discussion

A. Dataset Description

Chest X-rays are one of the most popular and cost-effective diagnostic imaging procedures utilized in medicine today. While interpreting a chest X-ray can sometimes prove challenging, this difficulty increases significantly when attempting to use machine learning algorithms or computer-aided detection and diagnosis (CAD). The lack of accessibility to extensive annotated data sets has made it difficult to develop CAD systems that could work effectively in actual healthcare settings utilizing chest X-rays. Creating substantial X-ray database collections requires an abundance of resources to label such vast amounts of images accurately. To address this issue, the National Institutes of Health (NIH) assembled the Chest X-ray Dataset [10], which comprises approximately 112,000 X-ray images bearing diagnoses from over 30,000 unique patients. A sample illustration of these illnesses can be seen in Figure 8.2.

According to the NIH dataset, the study divides the comprehensive dataset into three subsets to prepare and assess various machine learning models. These subsets are based on the proportion of patients' data represented within each category. Specifically the following:

* A training set makes up approximately 85% or 95,466 images of the overall dataset. This portion will be utilized to educate the model using supervised learning techniques.

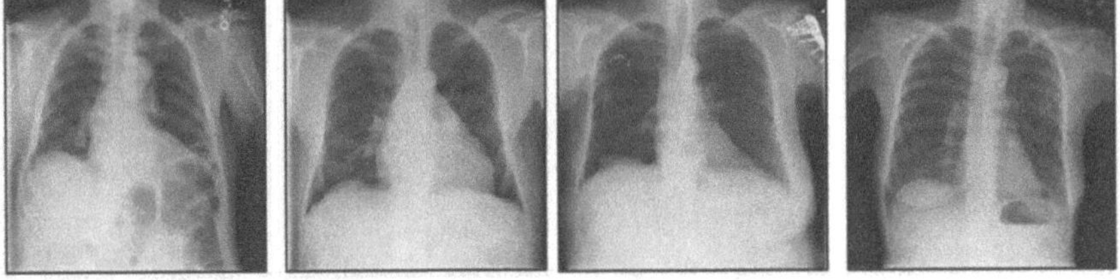

(**a**) Atelectasis (**b**) Cardiomegaly (**c**) Hernia (**d**) Infiltration

Figure 8.2 Sample chest x-ray image from the NIH-chest x-ray dataset.

* Approximately 10% (or 11,265) of the entire dataset represents the verification dataset. This subset serves as an internal evaluation tool during model development to ensure that performance metrics stay consistent across different segments of the data.
* The final 5% (or 5,389 images), known as the test dataset, serve as the final evaluative metric after all other checks have been completed. It offers unbiased results without any prior exposure to assist in predicting the model's general performance when confronted with new, unseen data.

B. Proposed Model Prediction

The study utilizes a deep learning model that processes 320 × 320 pixel X-ray images to identify various diseases, such as cardiomegaly, edema, and mass, with high accuracy. An example image is provided to illustrate the process. According to the ROC curve displayed at the bottom of the article, the model achieved impressive performance, particularly for diseases like cardiomegaly (AUC = 0.9), edema (AUC = 0.86), and mass (AUC = 0.82). While these results are encouraging, there is still room for improvement, ideally aiming toward a perfect score of 1. The authors also share findings from the GraphChxNet paper and involve radiologists in evaluating the model's output, which demonstrates its potential to accurately diagnose X-ray images and highlight the areas of focus when making predictions. Additionally, the model not only provides the primary diagnosis but also identifies secondary concerns, similar to what a radiologist would do during their examination, suggesting that the predictions may be influenced by factors such as X-ray noise or accumulated pathologies.

According to Figure 8.3, the model accurately predicted two conditions related to cardiac function: cardiomegaly (enlargement of the heart) and the absence of any visible masses or swelling in the chest cavity. However, the data also suggest that there is a higher likelihood of detecting masses

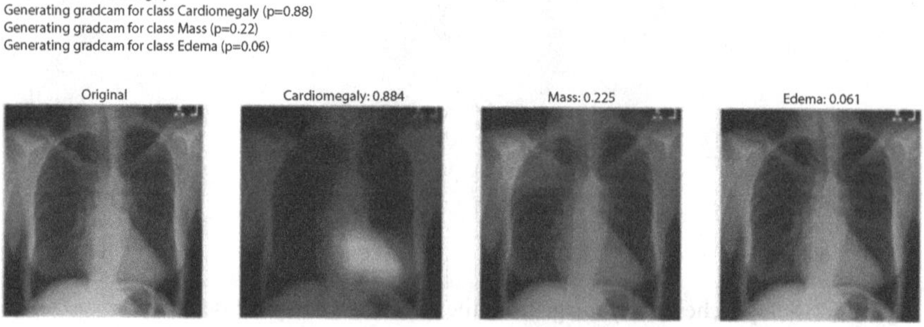

Figure 8.3 Model-predicted class—cardiomegaly.

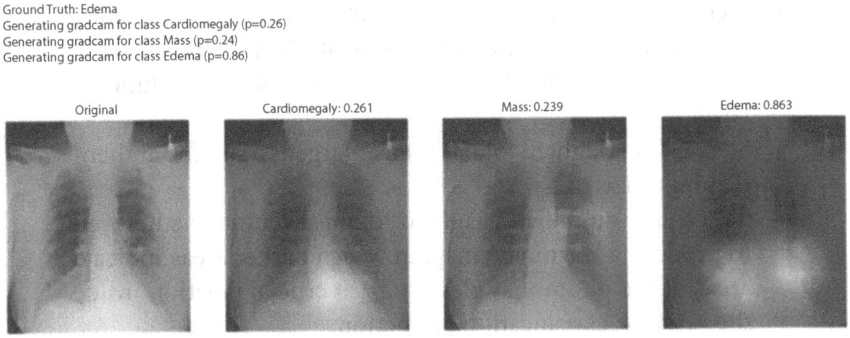

Figure 8.4 Model-predicted class—edema.

within the chest area, which could potentially be influenced by the shape of structures located near the center of the chest cavity.

According to Figure 8.4, the machine learning model identifies abnormalities in the chest region, particularly around the right side of the thorax. The model correctly detects fluid accumulation toward the lower part of the chest cavity. Interestingly, in spite of being absent in actual-world statistics, the version assigns a high rating to the label cardiomegaly for this photograph. This ought to offer treasured insights throughout mistake evaluation, which includes examining how the version focuses its attention on areas in which it makes predictions [11–13].

C. Data Visualization Using CAM

Classification Activation Mapping (CAM) is a treasured tool for understanding how deep neural networks make alternatives while classifying images. By highlighting areas within a photograph that contributes most significantly to its class final results, CAMs offer insights into which capabilities or components of the picture are crucial for accurate prediction. Moreover, these techniques enable us to research any capacity biases within the models themselves. In scientific imaging programs, like chest X-ray evaluation, CAMs have proven especially useful in figuring out the precise regions of the lungs suffering from illnesses, thereby enhancing diagnostic accuracy. To create a laptop-aided detection (CAD) version for a chest X-ray image, a series of common steps are followed. These include the following:

- Step 1: The photograph is, first of all, analyzed using an advanced device learning algorithm that is knowledgeable in distinguishing numerous types of chest X-rays, which include images of normals, pneumonias, atelectases, and others.

- Step 2: The rules create an output, which is subsequently used to compute a weighted sum of all the characteristic mappings generated by the framework's final convolutional layer.
- Step 3: The weighted total is then enlarged to suit the authentic photo size.
- Step 4: The amplified weighted total is eventually employed to produce a warmness map, in which darker areas indicate areas of the photo that play an extra crucial position inside the version's classification prediction.

The CAM (Computer-Aided Management) gadget can be superimposed onto the original X-ray photo to display visually the areas of the lung that are critical for the version's predictions. Radiologists may additionally use these data to beautify the accuracy of their diagnoses and inform remedy planning. For example, if the model identifies atelectasis in a chest X-ray photograph, the CAM can assist in pinpointing the correct regions of the lung affected by the infection. This information can then be used by the radiologist to verify the diagnosis and examine the quantity of the infection's severity.

In Figure 8.5, we present an example of a chest X-ray imaging study where various features have been identified using natural language processing techniques. The study includes three key components: the original image, correctly detected bounding boxes, and incorrectly detected false positives. Each of these components is represented by different colors to facilitate easy visualization. The green boundary marks the true locations

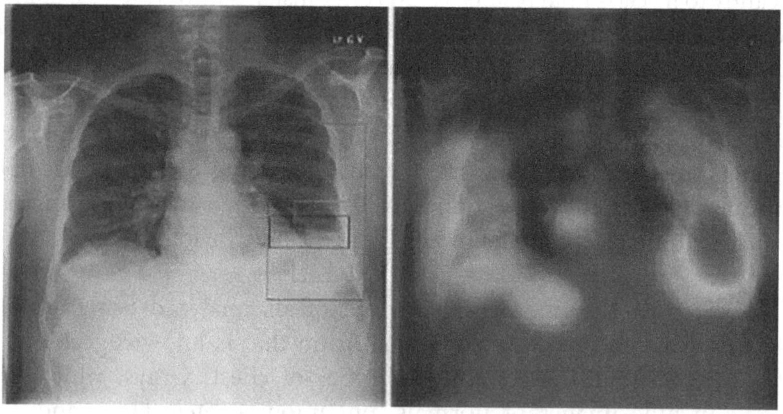

Figure 8.5 GradCAM visualization of class—atelectasis.

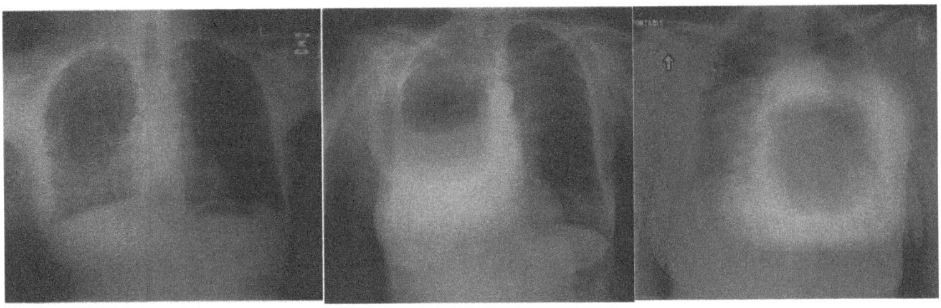

(a) (b) (c)

Figure 8.6 GradCAM visualization of class—pneumothorax, pleural fusion, cardiomegaly.

of the features, while the red ones indicate areas where errors occurred, and the blue lines represent the actual positions of the features based on expert annotations. By analyzing these patterns, healthcare professionals can gain valuable insights into the accuracy of their diagnoses and identify potential areas for improvement.

According to Figure 8.6(a), a patient exhibits a right-sided pneumothorax, which occurs when the lung collapses due to air or fluid accumulation in the pleural cavity. The image also depicts a chest tube inserted for treatment purposes. The AI version can hit upon abnormalities inside the lungs, together with pulmonary edema or pleural effusion, permitting healthcare vendors to quickly identify ability complications. For instance, in Figure 8.6(b), the version identifies an enormous right pleural fusion (fluid buildup within the pleural space) highlighting the affected location of the lung. Additionally, the model distinguishes among different forms of cardiovascular situations, which include congestive heart failure and enlarged hearts, as visible in Figure 8.6(c).

D. Evaluation Metrics

The performance of the proposed model is evaluated using area under the receiver operating characteristic (AUROC) curves and scores. A higher AUROC value shows a higher class, with a maximum value of 1 representing a great model. The AUROC curve plots the ratio of real positives to fake positives offering insights into the version's accuracy. The sensitivity, or powerful fee, is calculated as the share of real notable instances recognized by using the version among all exquisite instances inside the dataset. In essence, this degree reflects the version's capability to understand superb cases successfully.

$$\text{TPR} = \frac{True\,Positive}{(True\,Positives + False\,Negatives)} \tag{8.14}$$

The false-positive rate, or alternatively referred to as sensitivity, represents the proportion of overall negative instances within the dataset that the model incorrectly predicted as positive.

$$\text{FPR} = \frac{False\,positive}{False\,Positives + True\,Negatives} \tag{8.15}$$

The mathematical formula for calculating the AUC by ranking is as follows:

$$\text{AUC} = \frac{(TP + FN)}{(TP + TN + FP + FN)} \tag{8.16}$$

According to the table presented in this study (Table 8.1), the area under the receiver operating characteristic (AUROC) scores for each class were evaluated across different model variations. Additionally, the corresponding ROC curves are displayed in Figure 8.7 of the paper. These results indicate that the performance of the GraphChXNet-based model was superior compared to the other two models tested, thereby corroborating previous research findings reported in the scientific literature.

The network architecture of GraphChXNet is more advanced, enabling the model to glean more insights from data, potentially leading to better performance. GraphChXNet's design features a closer relationship between its layers, which helps minimize the number of model parameters and prevent overfitting. Additionally, the model incorporates batch normalization and skip connections to accelerate convergence and boost efficiency during training. Previously, we described how we utilized a pre-trained GraphChXNet model and modified its fully connected layers for our specific task. After evaluating the model's performance without additional training, we found that its ROC values were comparable to random guessing suggesting that its predictions were not particularly accurate. Nevertheless, we were able to use GraphChXNet as a fixed feature extractor and train an additional fully connected layer on top of it, resulting in improved performance. Throughout this process, we only updated the fully connected layers while maintaining the rest of the GraphChXNet architecture unchanged [14].

Table 8.1 AUROC scores for different model variants.

Pathology	ResNet (%)	EfficientNet (%)	CheXNet (%)	GraphChxNet (%)
Atelectasis	76.30	78.10	80.94	82.00
Cardiomegaly	74.10	88.40	92.48	91.20
Effusion	86.00	84.60	86.38	88.30
Infiltration	66.60	64.50	73.45	72.90
Mass	74.80	76.10	86.76	82.10
Nodule	57.20	62.40	78.02	72.00
Pneumonia	69.40	71.30	76.80	74.30
Pneumothorax	77.90	81.00	88.87	86.70
Consolidation	80.30	79.60	79.01	82.40
Edema	87.10	83.80	88.78	89.00
Emphysema	76.00	84.10	93.71	91.20
Fibrosis	68.70	73.10	80.47	80.80
Pleural thickening	71.50	71.20	80.62	74.70
Hernia	69.70	84.90	91.64	94.50

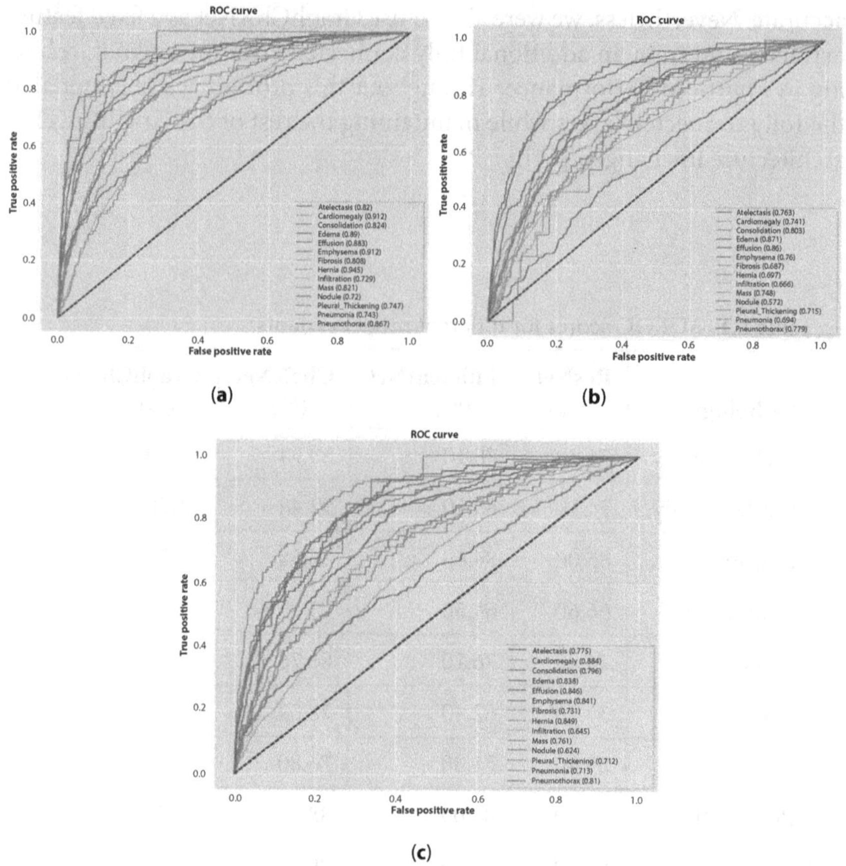

Figure 8.7 (a) AUROC curve for GraphChXNet, (b) AUROC curve for ResNet and (c) AUROC curve for Efficient Net.

8.4 Conclusion

Chest problems are among the leading causes of mortality globally, and early detection is crucial. Traditional manual diagnosis of lung conditions is time consuming and labor intensive, which highlights the need for efficient diagnostic tools. Our proposed automatic laptop-aided scientific machine can enhance radiological workflow by dashing up reading instances, lowering workload pressures, boosting productivity, and reducing healthcare fees. To reap these goals, we created a modern graph convolutional neural community (GCNN) model that is able to predict chest diseases with exceptional accuracy. We applied a progressive interest mechanism to assign adaptive weightings to embedded functions allowing the version to focus on the maximum applicable components of both node residences and topological systems. This method yielded awesome and mixed embeddings derived from node trends and topology. Through rigorous assessment and the use of the NIH Chest X-ray dataset, we verified that the GraphChXNet model extracts essential records from nodes and their relationships resulting in considerable upgrades in class accuracy. Our outcomes indicate that the proposed method surpasses mounted benchmarks throughout multiple overall performance metrics. These findings mean that the AI model developed herein is better equipped to become aware of severe chest abnormalities, thereby optimizing diagnosis and therapy processes. Additionally, Gradient-weighted Class Activation Mapping (Grad-CAM) became hired to visually represent the predicted anomalies in NIH photos and evaluate the model's potential to make informed decisions. Comparisons among Grad-CAM samples generated *via* the version those annotated by professional radiologists found out excellent agreement among the two units. Therefore, the GraphChXNet suggests commendable performance when judged against human professionals underscoring its capacity as a precious device in clinical imaging analysis.

References

1. Yu, X., Wang, S., Zhang, Y., CGNet: A graph-knowledge embedded convolutional neural network for detection of pneumonia. *Inf. Process. Manage.*, 58, 1, 102411, 2021.
2. Rajpurkar, P., Irvin, J., Zhu, K., Yang, B., Mehta, H., Duan, T., Ding, D.Y., Bagul, A., Langlotz, C.P., Shpanskaya, K., Lungren, M.P., Ng, A.Y., *CheXNet: Radiologist-Level Pneumonia Detection on Chest X-Rays with Deep Learning*, arXiv Cornell University, 2017, https://arxiv.org/pdf/1711.05225.pdf.
3. Meng, Z., Meng, L., Tomiyama, H., Pneumonia Diagnosis on Chest X-Rays with Machine Learning. *Procedia Comput. Sci.*, 187, 42–51, 2021. https://doi.org/10.1016/j.procs.2021.04.032.
4. Thekumparampil, K.K., Wang, C., Oh, S., Li, L., *Attention-based Graph neural network for semi-supervised learning*, arXiv Cornell University, 2018, https://doi.org/10.48550/arxiv.1803.03735.
5. Sriker, D., Greenspan, H., Goldberger, J., Class-Based Attention Mechanism for Chest Radiograph Multi-Label Categorization. *2022 IEEE 19th International Symposium on Biomedical Imaging (ISBI)*, Kolkata, India, pp. 1–5, 2022, doi: 10.1109/ISBI52829.2022.9761667.
6. Zhang, M., Wu, S., Gao, M., Jiang, X., Xu, K., Wang, L., Personalized Graph neural networks with attention mechanism for Session-Aware recommendation.

IEEE Trans. Knowl. Data Eng., 34, 8, 3946–3957, 2022. https://doi.org/10.1109/tkde.2020.3031329.

7. Bhatti, U.A., Tang, H., Wu, G., Marjan, S., Hussain, A., Deep Learning with Graph Convolutional Networks: An Overview and Latest Applications in Computational Intelligence. *Int. J. Intell. Syst.*, 2023, 1–28, 2023. https://doi.org/10.1155/2023/8342104.
8. Mudiyanselage, T.B., *et al.*, *Covid-19 Detection from Chest X-ray and Patient Metadata using Graph Convolutional Neural Networks*, ArXiv.org e-Print Archive, Cornell University, Ithaca, New York, 2021, ArXiv abs/2105.09720.
9. Chetoui, M., Akhloufi, M.A., Yousefi, B., Bouattane, E.M., Explainable COVID-19 Detection on Chest X-rays Using an End-to-End Deep Convolutional Neural Network Architecture. *Big Data Cognit. Comput.*, 5, 73, 2021. https://doi.org/10.3390/bdcc5040073.
10. Wang, X., Peng, Y., Lu, L., Lu, Z., Bagheri, M., Summers, R.M., ChestX-ray8: Hospital-scale Chest X-ray Database and Benchmarks on Weakly-Supervised Classification and Localization of Common Thorax Diseases. *IEEE CVPR*, 2017.
11. Batchuluun, G. and Choi, J., CAM-CAN: Class activation map-based categorical adversarial network. *Expert Syst. Appl.*, 222, 119582, July 15, 2023.
12. Bradley, A.P., The use of the area under the ROC curve in the evaluation of machine learning algorithms. *Pattern Recognit.*, 30, 7, 1145–1159, July 1997. https://doi.org/10.1016/s0031-3203(96)00142-2.
13. Basheer, S., Singh, K.U., Sharma, V., Bhatia, S., Pande, N., Kumar, A., A robust NIfTI image authentication framework to ensure reliable and safe diagnosis. *PeerJ Comput. Sci.*, 9, e1323, 2023.
14. Srivathsan, K., Bharath, S., Malini, A., *et al.*, Extended virtual reality-based memory enhancement model for autistic children using linear regression. *Int. J. Syst. Assur. Eng. Manage.*, 15, 58–67, 2024. https://doi.org/10.1007/s13198-023-02231-5.

9

Aspect-Based Sentiment Analysis Using GCN

Sachin K.*, Santhosh K.M.R., Sugindar A.D. and J. Felicia Lilian

Department of Computer Science and Business Systems, Thiagarajar College of Engineering, Madurai, Tamil Nadu, India

Abstract

A particular class of neural network designed specifically to function with graph-structured data is called a graph convolutional network (GCN). GCNs have gained popularity in recent years as a powerful tool for sentiment analysis of user-generated content on the internet. The rationale for this is because GCNs, which can take both the syntactic and semantic elements into account, are excellently adapted to aspect-based sentiment analysis (ABSA). They have been used to extract and analyze valuable information out from the vast volumes of text available, with positive results. This chapter offers a thorough examination of GCNs in ABSA highlighting the benefits, along with its drawbacks, of various methods and setting them within the larger framework of sentiment analysis (SA) using deep learning (DL).

***Keywords*:** ABSA, generative pre-defined transformer 3, graph convolutional network, NLP, visual question answering, graph neural network

9.1 Introduction

The fast-expanding discipline of natural language processing (NLP) focuses on examining how computers and human language interact. It includes the application of computer-based techniques for deciphering, understanding, and creating human language. NLP tasks comprise, but are not restricted to, speech recognition and machine translation. NLP made considerable

**Corresponding author*: sachinkamaraj.personal@gmail.com

Malini Alagarsamy, Rajesh Kumar Dhanaraj, J. Felicia Lilian, Vandana Sharma and Gheorghita Ghinea (eds.) Graph Convolutional Neural Networks for Computer Vision, (181–204) © 2026 Scrivener Publishing LLC

improvements in past years with the advent of DL. Evolution of DL led to mechanization of various manual works [28]. One crucial application of NLP is sentiment analysis, which seeks to ascertain the emotional tone behind a piece of text. This task is vital in natural language processing, as it enables machines to understand and interpret human emotions and opinions. Sentiment analysis is perhaps one of the means most used in several domains of application, focusing on customer service, marketing, and opinion mining. Due to the abundance of text data that are easily accessible by the public through the internet, sentiment analysis has been a powerful tool enabling businesses and organizations to understand public sentiment toward certain products, services, and brands. A detailed comparison of sentiment polarity in movie reviews was done by [1] in the year 2000, which is now an evolving field of research. There are two types of methodologies for performing traditional sentiment analysis as follows: lexicon based, where an analyst downloads a set of lexicons most commonly associated with positive and negative sentiments, and feature engineering strategies that involve a variation of decision trees, rule-based methods, and statistically generated decision tree-based models. However, the appearance of the new giant in the media landscape, ABSA, has delineated new opportunities in sentiment analysis. Generally, two of the most traditional approaches for implementing ABSA are machine learning and statistical techniques, though they present several disadvantages [2]. Consequently, even though ABSA aims at analyzing the sentiment within a text at an aspect-specific level, it uses deep learning methods, such as neural networks and the transformer model, to avoid the need to develop dictionaries of sentiment-indicating words. This helps in ascertaining the sentiment in a better way as it can factor in the vicinity of phrasing and other such entities. In addition, sarcastic and ironic content is easy to identify for the system, while traditional methods fail with it. Indeed, ABSA is on the rise in instances where it is used in analyzing sentiments, and it is expected to progress and become a lot better. Consider the illustration below.

"Although it is situated a little farther from the city centre, my experience during my stay in the hotel was great and I had a pleasant time. The room was tidy. At the restaurant, I also had supper, or a dinner as most of the restaurants of the western countries call it. While they might have been relatively slow with the service, the food however was very good."

In this case, the emotion toward the "The restaurant" at scenario 1 is intermediate, while at scenario 2, the emotion toward "Food" is positive while that toward "Service" is negative. While the sentiment analysis would

Table 9.1 Table illustrating the ABSA.

Area	Aspect	Sentiment
The hotel	Room	Positive
The hotel	Location	Negative
The restaurant	Food	Positive
The restaurant	Service	Negative

just allow one to determine whether the text was positive, negative, or neutral, the ABSA goes further to provide a deeper understanding of the text. Sentiment for each aspect, such as the room, location, food, and service, can be distinctly understood using aspect-based sentiment analysis, as illustrated in Table 9.1.

Among these models, there are transformers and neural networks used within the framework of the company's activities of ABSA. Other machine learning algorithms that have been implemented in ABSA include support vector machines (SVMs) and random forests, besides the neural networks and transformer models [30]. This ability makes it possible for these models to interpret non-linear and multiple path dependencies between inputs and outputs, a factor that makes them the best suited for analyzing the subject sentiment in text. Recurrent neural network (RNN), which is another model commonly employed in abstract-based sentiment analysis, can model convolutional features and temporal dynamics.

In recent years, it has become increasingly difficult to assess and gain meaningful analytic outcomes of text information on the internet because of textual information's year-on-year expansion. This has been solved by graph convolutional network (GCN). To represent and uncover the relations in the graph data, which is more effective for understanding the relations between the words in the phrase, we employed a type of neural network called GCN.

For the purpose of scoring the graph formulating data, certain neural network architectures called GCNs can be utilized. GCNs have been used lately in ABSA, and studies show they are efficient in many applications such as picture or text classification. The first step of GCN-based ABSA involves converting all the textual phrases in the review text into nodes in a graph, while the connection between each pair of phrases is depicted as the edges. In the context of sentiment analysis, there are a variety of approaches, with one such powerful model for analyzing text data being graph convolutional networks (GCNs).

The abovementioned networks are intended to be sensitive to the connections and context that exist in the text, which would be useful in emerging a deeper consideration on the undertone's underlying tenet. Furthermore, as GCNs are capable of processing huge quantities of data, they are useful for frameworks that call for paradigms related to the array and interconnection of ideas and text. Additionally, studies have found that GCNs preserve the strength of putting forth the context and semantics of a textual material, which is very important when isolating sentiments. The integration of GCNs into sentiment analysis has provided a chance to obtain a further view on the sentiment present within the text; it is desired that this will awaken more interest in the future.

Recent trends and research state the possibility and effectiveness of integrating GCN with ABSA, which mentions the evolution of a fine-tuned ABSA model with the incorporation of graphical structures with GCN, which plays a foremost task and boosts the overall execution when compared to existing models such as BERT and other language models. The core advantageous feature of GCN in ABSA is the representation of the context and relational link between them in a graphical way with weights on each link denoting the level of dependency between the various nodes it is connected to. The weights ease the bridging of appropriate words, and the difficult aspect is the initialization of weights. Erroneous initialization is one of the root causes of inaccurate results, and to avoid this, the model is being trained iteratively with weights being varied accordingly to provide a positive outcome until they reach a maximum level of satisfactory accuracy. GCN with ABSA overcomes the difficulties faced by traditional models, and notable features include faster performance with weights and high reasoning capability. Even though performance is improved, there are a few drawbacks in implementing sentimental analysis for enormous word counts, as constructing the graphical networks becomes difficult. As the field is currently under research and booming, these drawbacks can be overcome with further refinement, and there is promising need for the integration of GCN-ABSA with quantum computing given their high parallel processing power and large memory, which has the potential of overcoming the abovementioned issue.

To strengthen the accuracy and soundness of the model, visual question answering (VQA) can employ GCN and ABSA. GCNs can capture relationships between different entities in an image well, and ABSA helps to understand the emotion that pertains to a particular entity within an image. To make sense of this issue from the perspective of some VQA model, we may consider integrating these approaches with each other so as to better assimilate the contextual dimensions of any given photograph, which it would then use to generate more informative responses to queries. The main goal of visual question answering (VQA) at the intersection of CV and NLP is responding to queries about images using a natural language. ABSA plus GCN methods may be employed to enhance VQA's precision. When understanding what emotion connects with certain objects within a picture, it is useful to use ABSA, whereas through attributes, such as GCN, they can detect item–item relationships in an image. Therefore, integrating these techniques together helps make a VQA model that understands an image context more completely and supplies more relevant and detailed answers to questions.

Overall, GCN-powered sentiment analysis provides a pathway to explore massive datasets for useful insights. It is capable of improving the accuracy of sentiment analysis determination and provides users with a much clearer picture of the emotional state within the language. As for VQA, ABSA can be integrated with GCN to make the model more stable and precise when it has to make a decision. In this chapter, the author shall discuss the various methods used to achieve the results obtained, the outcomes of the research, and its possible future implications in this vast area of research.

9.2 GCN and ABSA

The process of extracting and evaluating large amounts of data in ABSA has become more undemanding due to the implementation of GCN. It has proven to be beneficial in a wide range of activities due to their competence in aligning semantics of the aspects.

9.2.1 The Meaning of ABSA

ABSA, an NLP task whose primary goals are to identify and extract information, namely, sentiments that may be expressed concerning one facet of an entity, has come a long way since its conception and is still an area of research and innovation.

The concept of ABSA was born with a need to analyze product reviews. Before the introduction of ABSA, traditional techniques, like rule-based systems, lexicon-based methods, and sentiment lexicons, were employed to perform sentiment analysis. The main disadvantage of using these traditional techniques were they were unable to handle intricate and subtle emotions, which in turn contributed to their reduced accuracy. Researchers, as a way to overcome this limitation, started to explore other techniques to extract more intricate, granular information from the text. This led to the formation of a new method in 2002 by [3], which has been acknowledged as the ABSA filtering technique, though it was at an early stage of its development. Pang and Lee also presented a categorization model of movie reviews based on good, bad, and neutral impressions conveyed on precise movie attributes, which they defined as plot, acting, and direction, among others. This marked a point of transition where the sentiment analysis began focusing more on the aspects of the text.

Midway through the 2000s, machine learning approaches for ABSA were explored by the researchers to improve accuracy. With the help of supervised learning algorithms, the categorization of text into their respective sentiment categories was proposed. Although this proposed method increased the accuracy of the SA, it was still limited by the availability of labeled data that were required for training the model and also by its inefficiency in handling complex and nuanced sentiments.

Toward the end of the 2000s, neural networks and deep learning techniques were explored by researchers for ABSA. To identify the attributes of the text and categorize the sentiments, CNNs and RNNs were proposed, which used unsupervised and semi-supervised learning techniques. Despite the increase in the accuracy and generalizability of ABSA, it still suffered handling complex and nuanced sentiments. The biggest issue in ABSA was that it required a thorough comprehension of the underlying subjects for the identification of aspects from the texts. This led to the proposal of unsupervised methods like topic modeling and clustering to identify aspects from the texts, which were more effective in the identification of the latent aspects of the texts. Still, these techniques required a sizable amount of labeled data to train the model and may have been ineffective for low-resources languages.

In a recent study, proposed a latent Dirichlet allocation (LDA)-based unsupervised aspect extraction approach [4], which was able to achieve an F1 score of 0.67. This was equivalent to the performance of supervised methods. In a different study, a neural topic model-based unsupervised technique was proposed by [5], which achieved an F1 score of 0.78 on a restaurant review dataset. Another technique for extracting aspects from the texts was by using rule-based methods, which depend on patterns or keywords of text. Despite being straightforward to use, along with increased customizations for other domains, it was still less reliable and precise when compared to other approaches. It was reported that an accuracy of 72.8%

was achieved by a rule-based approach in a dataset of product evaluations [6].

This built up to the advancements made in deep learning which subsequently led to the creation of ABSA. RNN and CNN have been greatly beneficial in enhancing the efficiency of absorption in ABSA. [9] used a CNN-based model, which was found to support an F1 score of 0.81 in a restaurant review dataset. This was a substantial improvement over the existing conventional techniques. In another research, [8] stated that an RNN-based model was able to achieve an F1 score of 0.87 on a laptop review dataset.

The most recent trend in the arena of ABSA is the integration of pre-trained large language models such as BERT and GPT-3. They showed enhanced efficiency for new and unknown inputs due to the vast amounts of data used during training. BERT has been successful in improving performance on diverse language comprehension tasks, while a GPT-3 model achieved an F1 score of 0.89 on a dataset containing reviews about restaurants [10]. Another essential advancement in ABSA is the growing use of multimodal data, including images and videos, that has resulted into newer fields like audio and visual sentiment analysis. It has been reported that a multi-modal sentiment analysis method combining text, sound, and visual information attained an accuracy level of 85.2% on a collection of movie reviews [11].

GCN (graph convolutional network), which is a semi-supervised neural network, is being studied for use in ABSA. This improves the accuracy and generalization capability of ABSA, since it uses graph-structured data for the determination of the relationship between different words in a phrase. GCN was proven to be highly effective in handling difficult or nuanced sentiments, unlike the approaches proposed earlier. This technique is now being repurposed to improve the performance of ABSA across different tasks and datasets accordingly.

9.2.2 Meaning of GCN

The graph convolutional networks were derived from the framework of graph signal processing (GSP). In contrast to traditional signal processing, which operates on signals defined over regular grids, the GSP processes signals that are graph based. The GSP's purpose is to come up with graph operations and Fourier transforms that can be used to extract meaningful information from data structured in a graph.

During the early 2010s, researchers of GSP began investigating how neural networks could be used in graph-structured data analysis. This led to the development of graph neural network (GNN) and spectral graph convolutional network (SGCN), which aimed to modify CNN fundamental function on the graphs. As such, these basic techniques constituted the bedrock upon which GCNs were built, thereby extending convolution on graphs.

In 2016, a significant breakthrough in this field was Welling and Kipf's proposal of the GCN model [12] that gives us a simple and efficient algorithm to perform graph-structured data classification using semi-supervised methods. On the graph, they designated "convolution" as the process used to get local neighborhood information from it. This approach resulted in state-of-the-art results on different benchmarks. GCNs are built on top of CNNs, which are widely used for image processing. Each node in a network is viewed as a "pixel" in GCNs, and the "neighborhood" of each node is defined by the edges connecting other nodes. Through the application of graph convolutions, GCNs can derive information from immediate neighbors of individual nodes. These features may be used for a number of purposes that include, but are not limited to, categorizing nodes and graphs.

The original GCN model was proposed by Kipf and Welling. The initial implementation is made up of the following two layers: a fully connected layer and a graph convolutional layer. The authors used citation network dataset (Cora, Citeseer, and Pubmed) and showed that GCN performs better than other models in a semi-supervised node categorization. It also proved that GCN can generalize to explore areas of the graphs that were previously unexplored.

GCNs are also used in ABSA, which analyze the sentiment of text toward a specific feature of an object. Analyzing the diverse and intricate verbal expressions of sentiment poses a considerable challenge in ABSA. It has been demonstrated that this problem can be effectively solved with GCN by modeling connections among aspects as well as emotions within the text.

New architectures and methodologies have led to significant advancements in graph convolutional networks (GCNs) in recent years. For example, as one such type of graph convolutional network (GCN), a novel graph attention network (GAT) has been introduced. GATs make use of attention mechanisms in GCNs that enable the model to focus on selected areas of graphs during feature learning, which leads to more effective information propagation [13].

Table 9.2 Assessment of AspectGCN's impact compared to other models on the SemEval-2014 dataset.

Model	Accuracy
AspectGCN	87.5%
LSTM-CNN	86.5%
Attention LSTM [7]	85.9%
MemNet	85.7%

Accordingly, this enhances its robustness and expressivity making it outperform many other models across various tasks. GATs have been able to achieve even better results than GCNs, with an accuracy of 74.2% on the aspect-based sentiment analysis task. Another variation of GCNs is GCN-LSTM, which uses a combination of GCN and LSTM. This architecture obtained an accuracy of 76.4%.

The ABSA model based on GCNs was proposed in a recent paper by Wang *et al.* and achieved industry-leading performance on a wide range of benchmark datasets. Aspect GCN model on SemEval-2014, which used GCNs to simulate the interactions between aspects and sentiments in the text, had an accuracy of 87.5%. Comparative performance of AspectGCN against other deep learning models on the SemEval-2014 dataset is detailed in Table 9.2.

9.3 Advancements of GCN and ABSA over the Years

The expansion of ABSA and advancement of GCN have made the job of examining the emotion of the sentence easier. This advancement has made ABSA flourish in various fields and applications. This enables individuals to have a deep understanding of the product expressed through public review.

9.3.1 Early Stages of ABSA

Sentimental analysis mainly focuses on extracting the sentiment of a product from public reviews, which enables the producer to know the needs and improvements on the product. Initially, "Bag of Words" (BOW), a model based on ML, was applied to forecast the sentiment hidden within

Table 9.3 Comparison of performance of standard and enhanced BOW model in terms of metrics.

Metrics	Bag of words (BOW) model	Sentimental analysis of online papers (SAOOP)
Accuracy	61.80%	81.70%
Precision	83.40%	85.60%
Recall	56.00%	86.70%

a phrase at document level. The main intention of the model is to reduce task complexity by paying attention to words and their frequency rather than their order in the document. This created a lack of focus on semantic words which had a crucial role in the prediction of the sentiment; however, this will not be a big issue in the case of small and informal text. Although the BOW model is simplified and diversely practiced, the model faces the following two major drawbacks: more effort and time are required to evaluate the words using a huge lexicon, which has duplication and repetition of words and poor accuracy. The reason behind it is syntactic arrangement, grammar, and semantic words are not taken into consideration in the sentiment prediction of a sentence.

Professor DM El-Din proposed a model named Sentimental Analysis Of Online Papers (SAOOP), an updated version of the BOW model, which analyzes a scientific paper's review. Because the SAOOP model analyzes word by word, weight is used instead of frequency. The SAOOP model is more advantageous than the standard BOW model for the following reasons: allows unmanned construction of a mini lexicon, saves the effort and time required for searching a word, improves accuracy, and can handle bulky reviews. From the Table 9.3, we infer that the enhanced BOW model has increased accuracy by 20% and exhibits better results [14].

In 2011, Khan A and Baharudin S proposed a model built on certain rules and independent of the domain that segregates the sentence into objective and subjective categories from weblog comments and feedback. The model's intent is to extract useful data from customer reviews collected online for making a decision. The semantics of the subjective sentence, which is generated from SentiWordNet (WorldNet Database), are used to determine the phrase's polarity. The proposed model performs more effectively than other machine learning models with greater accuracy of 87% at the feedback level compared to 83% at the sentence level. This clearly shows that the model outperforms well at the feedback level than at the sentence level [15].

Over the period of time, technology gets advanced and complex. This technical development has enhanced the ways of collecting and segregating the review of a product more easily and effectively. This led to the generation of a vast amount of reviews from internet users, which made it a challenging task to analyze reviews. The task of analyzing enormous reviews led to the collaboration between natural language processing (NLP) and computational linguistics called automated opinion mining. This could be accomplished at the document, sentence, and aspect levels. Most previous research works were carried out at the document and sentence levels, so there is a need to explore sentimental analysis at the aspect level. In 2015, Shibily Joseph and Chinsha T C proposed a model centered on syntactic arrangement semantic dependency and sentimental polarity for achieving opinion mining at the aspect level [16]. Their work has been evaluated on restaurant reviews and resulted in an accuracy of 78.04%.

Nowadays, product owners are particular about marketing their products and meeting the needs of the customer. This could be achieved not only by sentimental analysis (SA) because traditional SA delivers only the sentimental polarity of a customer toward a product through reviews. But actually, we need to deliver the sentiment for an aspect. An aspect is an entity that represents the topic, feature, or category of a product. Each aspect in the sentence delivers a different sentiment through which the product can be improved in a better manner and meet the needs of the consumer.

9.3.2 Establishment of Various GCN Models into ABSA

In recent years, ABSA has been implemented in the business sector by various projects, which include, but are not limited to, Bi-directional Encoder Representation Transformer and GPT. These models can identify only the aspect and their sentiment polarity for the sentence, and later on, attention-based neural networks have evolved, which made an efficient approach by focusing only on the specific context of the sentence for a specific aspect. Attention based approaches on neural networks bring out finer results than the model's without attention mechanism, and this has been manifested by many theses [26]. One of the theses proposes two neural networks, with and without attention mechanisms, to develop a question-answering chatbot independent of domain, and the outcomes show the superior nature of the model with attention mechanism. Another thesis says that the attention mechanism has upgraded the task of retrieving the accurate solution for a given query irrespective of the domain [27]. This attention-based approach can achieve sentimental analysis for a given aspect, but in some cases, there is a need for attention on the sentimental dependency between the contextual words and the aspect, which performs a significant role in improving SA. This has brought about the partnership of ABSA with GCN. GCN is a branch of deep structured learning that visualizes a group of data in a graph. The data are considered nodes and are interconnected *via* edges. These edges introduced the knowledge and importance of semantic dependency in ABSA through GCN.

Recent proposed neural network models lack sentimental dependency between words, which is due to sequential representation of the sentence. Professors B Huang and KM Carley proposed the target-dependent graph attention network (TD-GAT), also called syntax-aware ABSA, as it focuses on syntax structure and representation of the sentence graphically. This model represents the sentence in a dependency graph, which connects the target aspect and their related words. This model makes use of the concept of GAT, which is a variant of GCN, and its key component that generates aspect nodes from the sentence. The model created two variants, namely, TD-GAT with BERT and TD-GAT with GloVe, and both variants were evaluated on the SemEval 2014 dataset. The result of the evaluation says that TD-GAT with BERT is more efficient than TD-GAT with GloVe.

Aspect-specific graph convolutional network (ASGCN) models have resolved the problem of lack of a mechanism of ignoring words not associated with the specific aspect. ASGCN's primary objective is to properly utilize the syntactic dependency and long-range multi-word dependency within words through the dependency tree. The LSTM and masking mechanism were used to perform extreme filtration of features related to the target aspect through multi-layered graph convolutional networks. The model has introduced two variants, namely, ASGC-DG (for dependency graphs) and ASGCN-DT (for dependency tree). Technical and conceptual development of the sentimental relationships of words led to the establishment of various variants of GCN models to achieve ABSA. A detailed comparison of the various GCN-based models and their respective performances across multiple SemEval datasets is presented in Table 9.4.

9.3.3 Table Consisting of Details of ABSA and Technology They Were Built Upon

Table 9.4 Comparison of performance of various GCN models performed on SemEval—2014, 2015, 2016 datasets in terms of metrics.

Year of publication	Name of the model	Algorithm/ technique used	Variants	Metrics and scores							
				Laptop' 14		Restaurant' 14		Restaurant' 15		Restaurant' 16	
				Acc	F1	Acc	F1	Acc	F1	Acc	F1
Nov. 2019	ASGCN [19]	LSTM + GCN + masking mechanism	ASGCN-DT	74.14	69.24	80.86	72.19	79.34	60.78	88.69	66.64
			ASGCN-DG	75.55	71.05	80.77	72.02	79.89	61.89	88.99	67.48
April 2020	Relational graph attention network [20]	R-GAT + BERT	R - GAT	77.42	73.76	83.3	76.08	80.83	64.17	88.92	70.89
			R-GAT + BERT	78.21	74.07	86.6	81.35	83.22	69.73	89.71	76.62

(Continued)

Table 9.4 Comparison of performance of various GCN models performed on SemEval—2014, 2015, 2016 datasets in terms of metrics. (*Continued*)

Year of publication	Name of the model	Algorithm/ technique used	Variants	Metrics and scores							
				Laptop' 14		Restaurant' 14		Restaurant' 15		Restaurant' 16	
				Acc	F1	Acc	F1	Acc	F1	Acc	F1
Feb. 2021	SA-GCN [21]	BERT + SA-GCN + Top-k selection mechanism	SA-GCN + BERT	80.31 ± 0.47	76.99 ± 0.59	86.16 ± 0.23	80.54 ± 0.38	84.18 ± 0.29	69.42 ± 0.81	91.41 ± 0.39	80.39 ± 0.93
Oct. 2021	Sentic GCN [22]	SenticNet + LSTM	Sentic GCN	77.9	74.71	84.03	75.38	82.84	67.32	90.88	75.91
			Sentic GCN + BERT	82.12	79.05	86.92	81.03	85.32	71.28	91.97	79.56
Aug. 2021	DualGCN [23]	SynGCN + SemGCN	DualGCN	78.48	74.74	84.27	78.08	–	–	–	–
			DualGCN + BERT	81.8	78.1	87.13	81.16	–	–	–	–

(*Continued*)

Table 9.4 Comparison of performance of various GCN models performed on SemEval—2014, 2015, 2016 datasets in terms of metrics. (*Continued*)

Year of publication	Name of the model	Algorithm/ technique used	Variants	Metrics and scores							
				Laptop' 14		Restaurant' 14		Restaurant' 15		Restaurant' 16	
				Acc	F1	Acc	F1	Acc	F1	Acc	F1
Nov. 2022	KDGCN [24]	SynGCN + SemGCN + SenticNet	KDGCN	79	75.03	84.91	78.48	82.1	67.13	90.74	73.46
			KDGCN + BERT	82.6	79.55	87.23	81.69	85.98	72.4	93.66	82.49
Dec. 2022	MFLGCN [25]		MFLGCN	76.33	72.57	83.25	74.54	_	_	_	_

ASGCN, Aspect-Specific Graph Convolutional network; SA-GCN, Selective Attention-Based Graph Convolutional Network; DualGCN, Dual Graph Convolutional Network; KDGCN, Knowledge-Enhanced Dual-Channel Graph Convolutional Network; MFLGCN, Matrix Fusion-Based Graph Convolution Network.

9.4 Advancement of Technology with GCN and Algorithm Used

The development of increasingly advanced GCN algorithms, like graph attention Networks (GATs) and GraphSAGE, has made it possible to significantly enhance the performance of GCNs in a variety of applications.

9.4.1 Algorithm

An effective type of neural network, called GCNs, was created expressly to handle graph-structured data. The core algorithm used in GCN is graph convolution, which is an extension of traditional CNNs to work with graph-based data. The fundamental concept underlying GCN is to develop a graph convolution operation that enables the aggregation of data from a node's surrounding area. The algorithm iteratively carries out the convolution operations on every node updating each node's representation with information from its neighbors. Two major tasks of the GCN algorithm are as follows: linear transformation and non-linear activation.

Linear transformation is the very first step in the GCN algorithm. In linear transformation, the node of the feature matrix inputs undergoes a linear transformation, which is conducted using a weight matrix. It is similar to the traditional CNN, where the input data are filtered using a weighted matrix in the convolutional layer, and the output is passed to the next layer. A nonlinear activation function, such as sigmoid or ReLU, is then applied to the converted matrix similar to an activation layer in a traditional CNN. Such a nonlinear function makes the GCN learn much more complex and expressive data representations.

To aggregate data from various graph levels, the GCN algorithm can potentially be expanded to encompass many layers. A multi-layer GCN is what this is, and it can be used to model more intricate relationships in the data. For both semantic and non-semantic networks, performance rises as the network depth increases. Semantic models perform far better than models without semantics with the same amount of CNN layers. It is more efficient to feed the models with explicit semantics. The requirement for deeper networks to explore the implicit semantics is diminished by explicit semantics. In Table 9.5, we can conclude that CNN (one layer) GCN (one layer) excels perform better than CNN (two layers) [17]. To further enhance its performance, GCN can be integrated with other neural network architectures such as attention mechanisms.

Table 9.5 Comparison of the accuracy (%) and quantity of parameters between GCN and multilayer CNN.

Models	**Accuracy**	
CNN (one layer) GCN (one layer)	85.8	
	Model with semantic	Model without semantic
CNN (one layer)	84.1	82.2
CNN (two layers)	84.6	84.1
CNN (three layers)	85	84.8
CNN (four layers)	85.3	84.9
CNN (five layers)	85.5	85

9.4.2 Advancement of Technology Using GCN

Graph convolutional networks have been a key driver of recent technological advancements enabling more efficient and effective processing of graph-structured data.

GCN has been used for a number of applications, including collective classification and link prediction among others. Some of the real-world applications are traffic prediction, recommender systems, and drug discovery.

The process of node classification using a Graph Convolutional Network (GCN) is shown in Figure 9.1, where the input graph is preprocessed before being subjected to adjacency matrices and feature extraction to provide

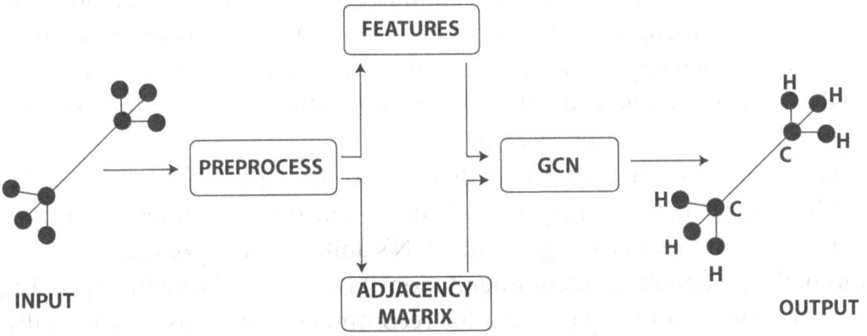

Figure 9.1 Node classification using GCN.

meaningful node labels. In node classification, GCN is used to assign labels to nodes in a graph, such as classifying nodes in a social network as individuals or organizations. In the prediction of a link, missing edges in a graph are predicted using GCN, such as predicting potential connections between individuals in a social network. In collective classification, GCN is used to classify all nodes in a graph collectively, such as classifying all nodes in a protein–protein interaction network as enzymes or receptors.

The combination of linear transformation and non-linear activation allows GCN to learn expressive representations of the data, and the ability to extend the algorithm to multiple layers and combine it with other architectures allows for even greater flexibility and performance. GCN has been effectively used for a variety of tasks, and it is expected that it will be a crucial tool for handling graph-structured data in the future.

GCNs are a subset of neural networks that are effective at processing data that are presented as a graph. This idea has facilitated growth in a number of fields, including CV and NLP, among others. The capacity of GCNs to accurately depict the complex connections between graph nodes is one of their important characteristics. This is achieved by utilizing convolutional operations on the graph structure, instead of the traditional grid-like structure of image data. GCNs are mainly applicable to problems where the data are highly interconnected and have a non-Euclidean structure.

In the fast-growing technology, collection and processing of data are also becoming complex. So, in this context, processing such data requires efficient methodology and techniques by which it can be segregated and utilized. As GCN is based on graph-based processing, it is able to process large-scale, complex graphs with high accuracy. Some of the domains where GCN is utilized are natural language processing (aspect-based sentiment analysis), Sociology, Marketing, Bioinformatics, and many more.

GCN possesses an advantage to process graph-structured data in a more efficient and effective way. Using GCNs, a huge spectrum of machine learning tasks may be enhanced to develop more meaningful representations of the data by utilizing the inherent structure of a graph, thus resulting in an increase in efficiency and accuracy in areas such as aspect-based sentiment analysis (NLP), image and video processing, natural language understanding, and recommendation system.

GCNs have been used to predict traffic patterns and identify traffic bottlenecks, which can help city planners and traffic engineers optimize traffic flow and reduce congestion. GCN's ability to analyze user behavior and make personalized recommendations take a special role in improving user experience and engagement for recommender systems. Additionally, ABSA with GCN is much more efficient at analyzing user sentiment.

In NLP, it has been applied in the optimization of neural language model efficiency. To enhance the understanding of a language, GCNs have, for example, been used in the encoding of syntactic and semantic information in phrases. It has also been used in storing relationships between words and sentences for many languages; it improves the effectiveness of the machine translation models. It has also helped in enhancing the sentimental analysis efficacy by considering various factors. Whereas the other models mostly rely on one feature to ascertain a sentence's polarity, the GCN model can be used to determine polarity based on a number of aspects of a single sentence.

CV employed GCNs to enhance the results of its semantic segmentation and object detection models. For example, using GCNs has been applied to better detect objects through the capture of the connections between different parts of an image. GCNs are applied to increase the effectiveness of image segmentation models by recording associations between diverse pixels of a single image.

In summary, the use of GCNs has improved a variety of issues and results in outstanding development in CV, NLP, and recommendation systems; it also has practical uses, such as drug discovery, traffic prediction, and recommender systems. As research in GCN advances, we are sure to see more innovations of this powerful technology in the future.

9.5 Case Study on GCN Application: Recommendation Systems

One of the applications of graph convolutional networks (GCNs) is recommendation systems. It helps in evaluating the user–item interactions and predicts the items that a user would find interesting using GCN. For example, it can be used to examine a network of user–item interactions where the things and users are shown as nodes. The edges between the nodes depict action such as purchasing and viewing. After training, it can be used to determine and predict future interactions of the user.

Using the convolution filters, GCN trains itself with the complex relationships between users and items applying the filters to each user and its neighborhood nodes with the relationship between the nodes and the structure of the item. A fully connected layer is responsible for the transfer of the GCN's output for the prediction or the forecast of a particular item.

GCNs provide customized recommendations in real life using past user–item interaction data. Thus, they help in finding the pattern in user behavior.

Overall, GCNs are helpful in finding, predicting, and forecasting the kinds of items that a customer may like based on the user–item interactions. It can also spot the trends in user behavior, which makes the recommendation system updated and more personalized.

9.5.1 Before Arrival of GCN in the Field of Recommendation Systems

Before the arrival of GCN, various other traditional methods were used for the building of a recommendation system. One such method is collaborative filtering, which was highly popular. This method focused on the common features between the user and the items and then made a suggestion based on it. For example, if two players choose the same category of products, then they are likely to choose the other products similarly too. Collaborative filtering is divided into the following two categories: item based and user based. Whereas the latter searches for patterns among things, the former looks for trends among people when making recommendations.

Another popular method is content-based filtering. This method depends on its suggestions regarding the characteristics of items. For instance, a content-based recommendation system can suggest additional action movies to a user who enjoys watching them. One of the main purposes of content-based filtering is to suggest products that are similar to those that a consumer has already loved.

Matrix factorization is another technique that seeks to reduce the dimensions of a matrix's representations of users and items. It can then find latent properties that account for the observed interactions between users and objects.

There are also hybrid approaches that integrate many techniques to provide recommendations [29]. A hybrid approach might, for instance, employ content-based filtering to further improve the recommendations after collaborative filtering has produced the initial set.

The specific requirements of the application determine which method is best, and each of the aforementioned approaches has pros and cons of its own. However, these methods suffer greatly from replicating the complex relationships between users and objects. For example, they may find it difficult to handle data sparsity or the dynamic nature of customer preferences. Effectively analyzing and classifying graph-structured data enable GCNs to make more accurate recommendations.

9.5.2 After Arrival of GCN in the Field of Recommendation Systems

Recommendation systems have come a long way since graph convolutional networks were first developed. A GCN is a type of neural network that is used to classify and evaluate the data, which are represented as graphs. Because of this, it is highly useful in applications, such as the recommendation system, in which data have a complex non-Euclidean structure.

The GCN has an advantage in the application of recommendation systems as it has the ability to manage data sparsity well. As the GCN architecture is complex, it has the advantage of distributing information efficiently across the graph even if it is scarce.

The GCN has the ability to represent the dynamic nature of user preferences. The GCN has the flexibility in adapting to new information, which allows it to change with user input. In comparison to traditional recommendation systems, it includes static user profiles sometimes providing less accurate information.

By adding additional information, such as the network structure, to the recommendation process using GCNs, recommendation performance can be enhanced. GCNs, for instance, can be trained to consider interactions between users and products as well as between users and commodities. This could provide useful information for recommendation making.

GCNs also have been used to improve the recommendation performance in cold-start scenarios where there is a lack of information about a new item or user. By leveraging the graph structure, even in the absence of direct interaction data, GCNs can still propose things or users based on their similarity. The GCN encoder and decoder methodology that was suggested in the study is shown in Figure 9.2. First, the user–item prediction model's GCN deep encoder is constructed, and the hidden layers of the encoder are activated using the ReLu activation function. Neighborhood data and graph topological similarity measurements make up the input. The embedding vector for each node given to the model is the output that is created. Second, by decoding the embedding vectors, the decoder forecasts the similarity score for each user–item pair. Reassembling the edges using the embedding space data is the primary stage in the decoding process, which represents the user–item relationship in a new space [18].

Thus, GCNs have had a big influence on the recommendation system industry. They have proven to be successful in predicting dynamic user

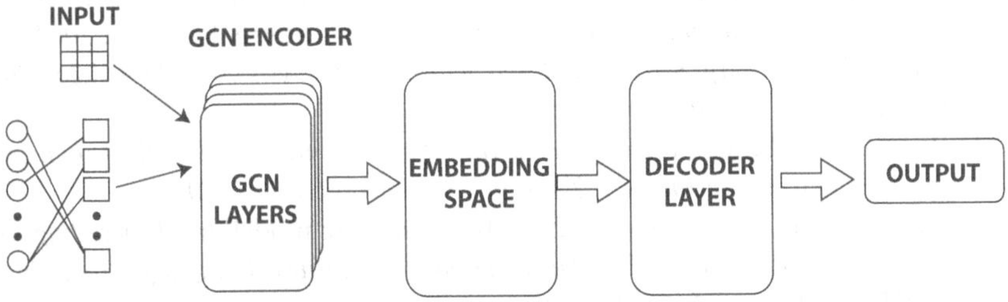

Figure 9.2 The steps involved in putting a GCN model into practice are as follows: creating graphs from data, designing a GCN, and optimizing forecast outcomes.

preferences, integrating extra data into the recommendation process, and managing the sparsity of data. In cold-start settings, GCNs have also been employed to enhance recommendation performance. Because of these benefits, GCNs are an effective tool in recommendation systems, and this area of study is still developing.

9.6 Summary

The practical uses of aspect-based sentiment analysis are numerous. If someone wants to know the overall tone of user or consumer reviews for a specific product or service, ABSA seems to be the best choice. ABSA is predicted to continue growing and improving in the future as it gains popularity. Because there is so much text data on the internet, it has become an essential tool for companies and organizations to assess customer sentiment about their products, services, and brands. GCNs handle a lot of data and are especially useful in understanding the connections between words in a phrase. Visual question answering (VQA) is one of the typical applications of natural language processing (NLP), which generates an answer from analyzing a visual content with a group of objects and their relationship between them. This combined progress of OpenCV and NLP has given good results and led to the advancement in NLP. Integration of GCN with VQA boosts the general performance, and in particular, graphs that connect objects make it easier to detect answer patterns. Sentiment analysis is a very exciting field within NLP and has the potential to significantly enhance our understanding of human language and emotions.

References

1. Pang, B. and Lee, L., A sentimental education: Sentiment analysis using subjectivity summarization based on minimum cuts, in: *Proc. 42nd Annu. Meeting Assoc. Computer. Linguistics (ACL)*, Stroudsburg, PA, USA, 2004.
2. Kumar, R. and Garg, S., *Aspect-Based Sentiment Analysis Using Deep Learning Convolutional Neural Network*, 2019.
3. Pang, B., Lee, L., Vaithyanathan, S., *Thumbs up? Sentiment classification using machine learning techniques*, 2002, *arXiv preprint cs/0205070*.
4. Li, X., Liu, Y., Liu, Y., Aspect extraction *via* topic modelling. *Proceedings of the 2018 Conference on Empirical Methods in Natural Language Processing-EMNLP 2018*, pp. 4166–4176, 2018.

5. Wu, Z., Li, X., Liu, Y., Neural topic models for aspect-based sentiment analysis. *Proceedings of the 2019 Conference of the North American Chapter of the Association for Computational Linguistics: Human Language Technologies*, vol. 1, pp. 3111–3120, 2019.
6. Hu, M. and Liu, B., Mining and summarising customer reviews. *Proceedings of the tenth ACM SIGKDD international conference on Knowledge discovery and data mining (KDD '04)*, Association for Computing Machinery, New York, NY, USA, pp. 168–177, 2004, https://doi.org/10.1145/1014052.1014073.
7. Wang, Y., Huang, M., Zhu, X., Zhao, L., Attention-based LSTM for Aspect-level Sentiment Classification, in: *Proceedings of the 2016 Conference on Empirical Methods in Natural Language Processing*, Association for Computational Linguistics, Austin, Texas, pp. 606–615, 2016.
8. Zhou, Z., Fan, Y., Liu, Y., Aspect-level sentiment classification with attention-over-attention networks. *IEEE Trans. Knowl. Data Eng.*, 31, 11, 2188–2199, 2019.
9. Devlin, J., *et al.*, *Bert: Pre-training of deep bidirectional transformers for language understanding*, 2018, arXiv preprint arXiv:1810.04805.
10. Sun, Y., Wang, X., Liu, Y., Fine-grained aspect-level sentiment analysis with pre-trained language models. *Proceedings of the 2019 Conference of the North American Chapter of the Association for Computational Linguistics: Human Language Technologies*, vol. 1, pp. 3100–3110, 2019.
11. You, Q., Li, W., Liu, T.Y., A multimodal approach to sentiment analysis of movies. *Proceedings of the 2015 IEEE International Conference on Multimedia and Expo-ICME 2015*, pp. 1–6, 2015.
12. Kipf, T.N. and Welling, M., *Semi-supervised classification with graph convolutional networks*, 2016, *arXiv preprint arXiv:1609.02907.*
13. Veličković, P., *et al.*, *Graph attention networks*, 2017, *arXiv preprint arXiv:1710.10903.*
14. El-Din, D.M., Enhancement Bag-of-Words Model for Solving the Challenges of Sentiment Analysis. *Int. J. Adv. Comput. Sci. Appl.*, 7, 1, 2016.
15. Khan, A. and Baharudin, B., *Sentiment classification using sentence-level semantic orientation of opinion terms from blogs*, IEEE, 2011.
16. Chinsha, T.C. and Joseph, S., A syntactic approach for aspect based opinion mining. *Proceedings of the 2015 IEEE 9th International Conference on Semantic Computing (IEEE ICSC 2015)*, Anaheim, CA, USA, 2015, pp. 24–31, doi: 10.1109/ICOSC.2015.7050774.
17. Zhang, P., Lan, C., Zeng, W., Xue, J., Zheng, N., *Semantics-Guided Neural Networks for Efficient Skeleton-Based Human Action Recognition*, Apr. 2, 2019, *arXiv:1904.01189v1 [cs.CV].*
18. Özcan, A., Applying graph convolution networks to recommender systems based on graph topology. *Dicle Univ. J. Eng. (DUJE)*, 13, 2, 205–211, June 28, 2022.

19. Zhang, C., Li, Q., Song, D., Aspect-based Sentiment Classification with Aspect-specific Graph Convolutional Networks, in: *Proceedings of the 2019 Conference on Empirical Methods in Natural Language Processing and the 9th International Joint Conference on Natural Language Processing (EMNLP-IJCNLP)*, Association for Computational Linguistics, Hong Kong, China, pp. 4568–4578, 2019.
20. Wang, K., Shen, W., Yang, Y., Quan, X., Wang, R., Relational graph attention network for aspect-based sentiment analysis, in: *Proceedings of the 58th Annual Meeting of the Association for Computational Linguistics*, pp. 3229–3238, 2020.
21. Hou, X., Huang, J., Wang, G., Huang, K., He, X., Zhou, B., *Selective attention based graph convolutional networks for aspect-level sentiment classification*, 2021, *arXiv preprint arXiv:1910.10857[V4]*.
22. Liang, B., Su, H., Gui, L., Cambria, E., Xu, R., Aspect-based sentiment analysis *via* affective knowledge enhanced graph convolutional networks. Elsevier. *Knowledge-Based Syst.*, 235, Article 107599, 2021.
23. Li, R., Chen, H., Feng, F., Ma, Z., Wang, X., Hovy, E., Dual Graph Convolutional Networks for Aspect-based Sentiment Analysis, in: *Proceedings of the 59th Annual Meeting of the Association for Computational Linguistics and the 11th International Joint Conference on Natural Language Processing (Volume 1: Long Papers)*, Online. Association for Computational Linguistics, pp. 6319–6329, 2021.
24. Zhang, Z., Ma, Z., Cai, S., Chen, J., Xue, Y., Knowledge-Enhanced Dual-Channel GCN for Aspect-Based Sentiment Analysis. *Mathematics*, 10, 4273, 2022.
25. Wang, J., Li, X., An, X., Modeling multiple latent information graph structures *via* graph convolutional network for aspect-based sentiment analysis. *Complex Intell. Syst.*, 9, 4003–4014, 2022.
26. Sundarakantham, K., Lilian, J.F., Rajashree, H., Shalinie, S.M., ACA: Attention-Based Context-Aware Answer Selection System, in: *Machine Intelligence and Signal Processing: Proceedings of International Conference, MISP 2019*, Springer, Singapore, pp. 327–338, 2020.
27. Lilian, J.F., Sundarakantham, K., Shalinie, S.M., QeCSO: Design of hybrid Cuckoo Search based Query expansion model for efficient information retrieval. *Sādhanā*, 46, 181, 2021.
28. Malini, A., Priyadharshini, P., Sabeena, S., An Automatic Assessment of Road Condition from Aerial Imagery Using Modified VGG Architecture in faster-RCNN Framework. *J. Intell. Fuzzy Syst.*, 40 11411–11422, Jan. 1, 2021.
29. Lilian, J.F., Vetriveeran, D., Malini, A., Kayalvizhi, S., HQA Bot: Hybrid AI Recommender Based Question Answering Chatbot. *Int. J. Intell. Syst. Appl. Eng.*, 11, 4, 227–33, Sep. 21, 2023.
30. Thirisha, R., Kiruthiga, A.S., Arunkumar, S., Lilian, J.F., SSATS—Enhancement of Semantic Similarity of Abstractive Text Summarization Using Transformer, in: *International Conference on Information Technology*, Springer Nature Singapore, Singapore, pp. 87–97, Mar. 2, 2023.

10

Analysis and Classification Using Graph Convolutional Neural Networks in Medical Imaging

M. Suguna[1] and Priya Thiagarajan[2]*

[1]Department of Information Technology, Thiagarajar College of Engineering, Madurai, TN, India
[2]Department of Computer Science and Business Systems, Thiagarajar College of Engineering, Madurai, TN, India

Abstract
***Background and Purpose*:** Convolutional neural networks (CNN) are commonly utilized to analyze and classify medical images. It is still limited in its ability to only consider spatially adjacent data in the Euclidean space to extract features from an image. This may lead to missing out any latent correlation present in the non-Euclidean feature space. Graph convolutional neural networks (GCNN) have a highly desirable property of having the flexibility to split an image into a set of nodes (each pixel or superpixel can be considered as a node) and edges and connect them together for processing in such a way that maximizes the detection of finer features. This is because when processing the data, not only are the local similarities considered but the nonlocal similarities are also utilized. This means the features are extracted from spatially adjacent data and also spatially distant data by the graph convolutional layers.
***Study Methodology*:** This study aims to investigate the advantages of GCNN over conventional CNN. It also discusses various applications of graph convolutional neural networks in image processing, with a focus on applications in ophthalmology. In ophthalmology, several diagnoses are made based on non-invasively captured images like the retinal fundus image (RFI) and optical coherence tomography (OCT). Eye disorders, including diabetic retinopathy (DR) and glaucoma; neurological conditions and diseases, like papilledema and Alzheimer's disease; and even systemic conditions, including renal and cardiovascular diseases are

**Corresponding author*: ptca@tce.edu

Malini Alagarsamy, Rajesh Kumar Dhanaraj, J. Felicia Lilian, Vandana Sharma and Gheorghita Ghinea (eds.) Graph Convolutional Neural Networks for Computer Vision, (205–224) © 2026 Scrivener Publishing LLC

diagnosed using the RFI. Similarly in radiological diagnoses, GCNN is seen to perform better with complex image analysis in published literature.

Also, GCNN can be used for denoising images before further processing or classification tasks.

We also experiment and present a system of medical image classification using GCNN to distinguish "normal," "papilledema," and "pseudopapilledema" in retinal fundus images.

Conclusion: Thus, this paper explores the possibility of using GCNN instead of, or in addition to, conventional CNN in the analysis and classification of medical images to exploit its superior performance for non-Euclidean feature extraction.

Paper Type: Book chapter

Keywords: CNN, GCNN, GCN, healthcare, image classification, non-euclidean

10.1 Introduction

10.1.1 Artificial Intelligence (AI) Applications in Healthcare

AI techniques in healthcare are presenting new ways to process, analyze, and use data. This can remarkably improve patient outcomes. AI applications are used to help healthcare experts make informed decisions, identify patients at risk, and develop personalized treatment plans [1, 2].

A few applications of AI in the healthcare domain are discussed here.

i) Disease detection by image classification: X-ray, MRI, and CT scan images analyzed by AI tools after suitable preprocessing. This can help extract finer features from the images to detect anomalies that may not be noticeable by human experts [3]. This can enhance disease diagnosis rates, particularly in the early stages.
ii) Predictive analytics: EHRs can identify high-risk individuals for diseases or disorders. They can be used to predict the success of therapies [4].
iii) Treatment options: Artificial intelligence (AI) tools help healthcare professionals make better selections. Using real-time data to analyze a patient's condition, drugs, and history can aid in selecting effective treatment alternatives [5].
iv) AI may analyze genetic data and patient-specific information to customize treatment approaches. These can be tailored to each individual's unique demands and genetics [6].

v) Virtual health assistants: Intelligent chatbots and voice assistants can aid patients in managing their treatment effectively. They can provide guidance on medication, scheduling appointments, and answering common health questions [7].

AI is proving its ability to positively impact healthcare applications by faster, more precise diagnoses and personalized treatments. There are also various obstacles to overcome before AI is extensively implemented in healthcare. They include data privacy, regulations, and ethical considerations. These challenges have to be addressed effectively and efficiently [8, 9].

10.1.2 Artificial Intelligence for Image Processing Tasks in Healthcare

Image processing, segmentation, and classification have improved significantly with the introduction of intelligent systems. AI algorithms can now analyze medical images (retinal fundus images, optical coherence tomography images, X-rays, CT scans, and MRIs) to diagnose diseases in various specialties including ophthalmology, cardiology, oncology, and neurology. AI tools and techniques can also help identify early signs of disease using finer features not visible to the human eye. This helps in earlier intervention leading to better patient outcomes [10].

AI can help radiologists interpret images more quickly and accurately reducing the time it takes to diagnose and treat patients [11, 12]. AI algorithms can be used for tedious processes, like segmentation, and to highlight any abnormalities. They can also be trained to suggest possible diagnoses enabling radiologists to concentrate on other critical areas. Usually, neural networks are used for these tasks.

AI algorithms are also used to enhance medical images making it more convenient for doctors to interpret them, especially in telemedicine applications. AI can reduce noise in images, adjust contrast and brightness, and remove unnecessary features [13, 14].

AI tools are also helpful for planning surgeries. They can help by analyzing medical images and creating 3D models of organs, tissues, and blood vessels. AI can also provide real-time feedback during surgical procedures helping surgeons understand and operate on complex anatomical structures [15].

AI algorithms can track changes in individual medical data over time. This will allow doctors to monitor the progression of diseases and the effectiveness of treatments with smart health monitoring systems [16].

AI in image processing in healthcare has increased efficiency and has a potential to reduce healthcare costs. But it is equally important to stress that AI algorithms cannot be considered a replacement for human expertise. They should be used by trained professionals in conjunction with clinical judgment. They are ideally suited for triaging and to assist healthcare specialists to holistically assess and help patients.

This chapter concentrates on the usage of AI for image analysis and, in particular, the applications using graph convolutional neural networks for image processing tasks. Recently, graph-based deep learning methods are also gaining popularity due to their effectiveness and success in various applications [17].

10.1.3 Concept of Neural Networks

A neural network is an AI model trained to make decisions similar to the human brain. It has several interconnected processing units called neurons. These units collaborate with each other to resolve complex and intricate problems.

Neural networks are the preferred networks for image processing tasks in healthcare [18].

The neurons in a neural network are organized into layers, with each layer performing a specific function. The input layer receives input data, usually image or text, and passes it to the next layer. After the data are processed by each of the hidden layers, the output layer generates the final output. This could be a prediction, classification, or decision.

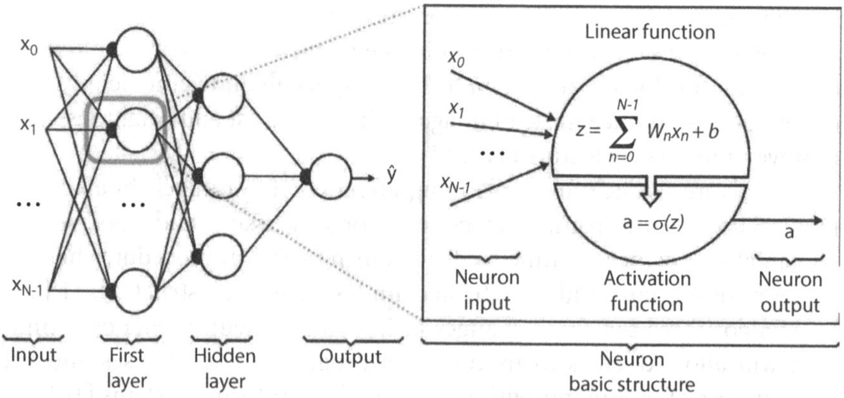

Figure 10.1 Basic neural network structure with three layers for n inputs. Image source: ref. [18].

Neural networks contain several hidden layers, which perform the majority of the processing tasks. Each neuron in a layer receives the outputs of neurons in the previous layer. It carries out specific mathematical operations on the received inputs and feeds the output to neurons in the next layer (Figure 10.1).

In the training process, the neural network adjusts the weights assigned to the connections between neurons to reduce the discrepancy between its predicted output and the actual output. This is performed by "backpropagation." Backpropagation is a technique that involves propagating the error backwards through the network. The connection weights are optimized to minimize the error function.

Neural networks are utilized for a variety of applications, including gesture detection, picture identification, speech recognition, language processing, and gaming.

10.1.4 Benefits of Neural Networks

Neural networks outperform traditional machine learning algorithms in many medical image processing applications. Medical imaging databases are huge and complex. Classic machine learning algorithms struggle to process and interpret these datasets. Neural networks can handle large, complex images and data [19]. They can efficiently extract features and patterns from images, which standard algorithms may struggle with. Neural networks can automatically extract data from medical images. This increases productivity and eliminates human error [20]. Traditional machine learning methods require manual computations for feature engineering and extraction. These calculations are complicated and time consuming. Manual feature extraction may miss some of the most significant image features.

Neural networks can achieve high accuracy in image analysis tasks. This is almost always true when trained on large, high-quality datasets. This makes neural networks highly suitable for healthcare tasks, such as disease diagnosis or monitoring [21].

Neural networks can process images much more quickly than human experts. This makes them ideally suited for real-time applications such as surgical assistance.

Neural networks can identify non-linear functions from the given data. This enables neural networks to capture complex relationships between the input features and output predictions. Traditional machine learning algorithms are linear and have limited ability to capture complex relationships. Neural networks also have the ability to accept and learn from new data.

This means that as new medical images become available, the neural network can learn from the new data. This will improve its performance over time.

Neural networks were often considered black boxes. The processing information of the decision making of the neural network is not readily available. Recent advances in explainable AI techniques have made it possible to understand and interpret the decisions made by neural networks. This is particularly important in medical image processing. The doctors will trust the diagnostic system better if they can understand how the algorithm arrived at its diagnosis or prediction [22].

Convolutional neural networks (CNNs) and its variants are the most preferred neural network models for image processing in healthcare [23]. In this chapter, we will discuss the applications of graph convolutional neural networks (GCNNs) for medical image processing in comparison with CNN.

Section 10.2 is a detailed literature review of the latest published literature using GCNNs for image processing in healthcare. In Section 10.3, we will present the methodology of our project to demonstrate the use of GCNN for image classification for retinal fundus images in ophthalmology. The results of our experimentation with CNN and GCNN are detailed in Section 10.4. Recommendations for future work, along with the conclusions of our study, are presented in Section 10.5.

10.2 Literature Review—GCNN in Healthcare

The literature review of this paper focuses on the use of GCNN in healthcare specialties where AI image processing is being employed successfully. Papers using GCNN for image processing and classification tasks are presented here.

[24] present a dual graph reasoning unit (DGRUnit) for brain tumor segmentation that employs GNNs to capture contextual and relational features. The DGRUnit is made up of the following two sections: a spatial reasoning section that models spatial links across tumor regions and a channel reasoning unit that aggregates semantic dependencies across channels. The DGRUnit outperforms several existing methods in brain tumor segmentation tasks.

The need for interpretable models in medical image analysis in histology. It proposes a graph-based approach to model cancer tissue histology using a graph of nuclei. It is analyzed using two novel graph convolutional network frameworks for disease classification and visualization. The proposed frameworks accurately distinguish between different types of breast and prostate cancers while highlighting important visual details [25].

A novel graph convolutional neural network, ResGNet-C. This system is designed to categorize lung CT images into either normal or indicative of COVID-19-related pneumonia. Two advanced models, namely, NNet-C and ResNet101-C, are derived as successful outcomes. Through five cross-validations on a dataset comprising 296 CT images, the approach attains high accuracy, sensitivity, and specificity. The suggested graph construction method is straightforward and adaptable. It is beneficial in enhancing classifier performance [26].

[27] discusses a SARS-Net model. It is an intelligent system that uses both GCNN and CNN together to detect COVID-19 infection in patients. This is done by analyzing features in chest X-ray images. The proposed model achieves high accuracy and sensitivity on the validation set, outperforming previously mentioned currently existing methods.

The drawbacks of other computer-assisted diagnosis systems for COVID-19 detection like limited datasets and performance measures. They have used a huge chest X-ray dataset curated from several sources and developed a supervisory mechanism. They also introduced a COVID-19 GCN, which uses data from multiple sites. This method efficiently determines disease development by using the dataset—the state of the training samples and beginning scores. These proposed methods demonstrate high efficacy and superior performance, particularly with GCN [28].

Zhu, Y., *et al.* have proposed an early Alzheimer's disease diagnosis system using a Graph Convolutional Network (GCN). It combines interpretable feature learning and dynamic graph learning. This has improved the effectiveness and interpretability of the GCN. The module of interpretable AI is used to select informative features while abandoning redundant ones. Then, the dynamic graph learning module adjusts neighborhood relationships to refine the classifier. Experiments show that the method achieves high levels of diagnosis performance and also provides interpretability for personalized disease diagnosis [29].

A re-weighted adversarial graph convolutional network (RA-GCN) for dealing with classes. This method uses a graph-based neural network for each class, varying the weights of class samples to change their importance to the classifier. Adversarial training optimizes the parameters of both the classifier and weighting networks. The paper demonstrates the superior performance of RA-GCN in comparison to recent methods across synthetic and publicly available medical datasets [17].

In the article of Chen, Z., *et al.* the significance of automated diagnosis for 3D medical data in the context of intelligent healthcare is discussed. The proposed system is an instance importance-aware graph convolutional network (IGCN). It is integrated with a multi-instance learning (MIL) framework. The system calculates the importance of each 2D slice and incorporates this information to refine the diagnosis branch. The IGCLayer is used to extract additional features in both significance and characteristic-based configurations. The challenge of insufficient supervision in 3D datasets is addressed by the importance-based sub-graph augmentation (SGA). Experimental findings on CC-CCII and PROSTATEx datasets confirm the better efficacy of the proposed method over other existing methods [30].

The DE-DCGCN-EE, a deep learning-based approach for retinal vascular segmentation. It is made up of a dual encoder to detect edges, a dynamic-channel GCN, and an edge enhancement block. This approach achieves more accurate segmentation of fine blood vessels. This is validated by experimental results on five retinal image datasets [31].

A framework for predicting diseases using multiple medical data types. The generalizable graph convolutional framework learns to create a population graph. which has variational edges. This improves the accuracy of disease prediction. Monte Carlo edge dropout uncertainty estimation is employed to assess predictive uncertainty. The experimental outcomes indicate that this method improves the accuracy of predicting autism spectrum disorder, Alzheimer's disease, and ocular diseases [32].

Wang *et al.* introduce a method for the detection of autism spectrum disorder (ASD) using fMRI data. This method employs multi-atlas graph convolutional networks and ensemble learning techniques. Data from 949 participants from the Autism Brain Imaging Data Exchange (ABIDE) has been used in this study. The proposed technique has an AUC of 0.8314 and an accuracy of 75.86%. This outperforms other methods published in recent years and shows promise for the accurate detection of ASD [33].

GCNNs have shown significant and promising results in some cases. There is still much work to be done to establish their effectiveness for various medical imaging tasks in healthcare. This will make them more acceptable for the healthcare professionals and can widely be used in clinical settings.

10.3 Methodology

A detailed discussion of the methodology of this study is presented here.

- Dataset selection
- Exploratory data analysis
- Preprocessing of data
- Image classification using GCNN
- Performance metrics of the proposed system

10.3.1 Dataset

The dataset we have chosen to use for demonstrating medical image classification is the "Machine Learning for Pseudopapilledema Dataset" from Open Science Framework (OSF) [34].

The dataset consists of three classes of images of the retinal fundus, "normal," "papilledema," and "pseudopapilledema"—779 normal retinal fundus images and 279 each of papilledema and pseudopapilledema. Papilledema and pseudopapilledema exhibit swelling of the optic nerve head (with minor variations in the characteristics), which is visible in the retinal fundus images.

10.3.2 Exploratory Data Analysis (EDA)

EDA of the images is performed on the data. Since it is an image dataset for classification without any additional data about the patients or any other demographic or comorbidity information, the quantity of images in every class is plotted. Data visualization is done to display a selection of images from every class.

10.3.3 Data Preprocessing and Image Augmentation

The exploratory data analysis reveals an imbalance in the quantity of images in each class. Data trimming is performed by using random samples of images from the "normal" class to keep the classes balanced in terms of the quantity of images in every class.

Data augmentation is performed by rotating or flipping the images. This is done randomly. This is done to make the system robust to different orientations of the retinal fundus images.

10.3.4 Image Classification Using Graph Convolutional Neural Networks

Most medical image classification systems use the vanilla CNN model or one of its variants. But the CNN models have the limitation of its ability to only consider spatially adjacent data in the Euclidean space to extract features from an image [35]. It is possible that this could result in overlooking any underlying correlation that exists within the non-Euclidean feature space.

Graph convolutional neural networks consider the image as a graph.

Here, we will consider the graph convolutional neural network (GCNN), which has flexibility to split an image into a set of nodes (each pixel or superpixel can be considered as a node) and edges and connect them together for processing in such a way that maximizes the detection of finer features. This is because, during data processing, not only are the local similarities considered but the nonlocal similarities are also utilized (Figure 10.2). This means that the features are extracted from spatially adjacent data and also spatially distant data by the graph convolutional layers.

Though both CNNs and GCNNs are used for image processing, they differ in the way they handle image data.

CNNs are mainly designed for processing 2D images and are extensively used in computer vision tasks. The convolutional layer in a CNN is used to extract features from the given images. Element-wise multiplication is done. This is performed by sliding a filter or kernel with suitable weights across the image pixels. The aggregated values are used to create a feature map. This is repeated with different filters, with different weights. This generates multiple feature maps. These are used to capture different features from the images. These feature maps are passed into the fully connected layers to perform classification or regression.

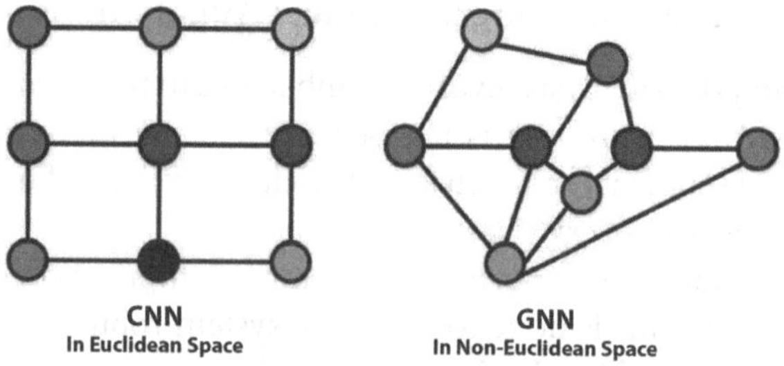

Figure 10.2 Concept of CNN and GCNN image analysis. Image source: ref. [35].

Graph-structured data are the application domain for GCNNs. In a graph, the data points are considered as the nodes. The connections or interactions between the nodes are shown by edges. In image processing, each pixel is a node. The edges represent the spatial relationships between the pixels. GCNNs are used to extract the image's global and local properties. This is performed by graph convolutional operations. They combine information from the node's spatial neighbors.

10.3.4.1 Convolutional Neural Network (CNN)

CNNs are used for image processing, segmentation, and classification. They are capable of learning spatial hierarchies of features from input images. CNNs can automatically learn to detect edges and corners in images. They use these features to recognize more complex patterns in the image.

Figure 10.3 illustrates the basic design of a CNN.

The initial layer is a convolutional layer. This extracts local characteristics by applying a set of filters of a given size to the input image. These are then processed through a non-linear activation function like the ReLU.

The subsequent layer is a pooling layer. This decreases the dimension of the feature maps by downsampling them. This is done to decrease overfitting and improve model robustness.

Several sets of these two layers are used, with each set of convolutional + pooling layers learning more complex features than the previous one. The result of the last convolutional layer is passed into one or more fully linked layers. They complete the classification or regression objective. The number of layers can be modified to our requirements for optimal performance.

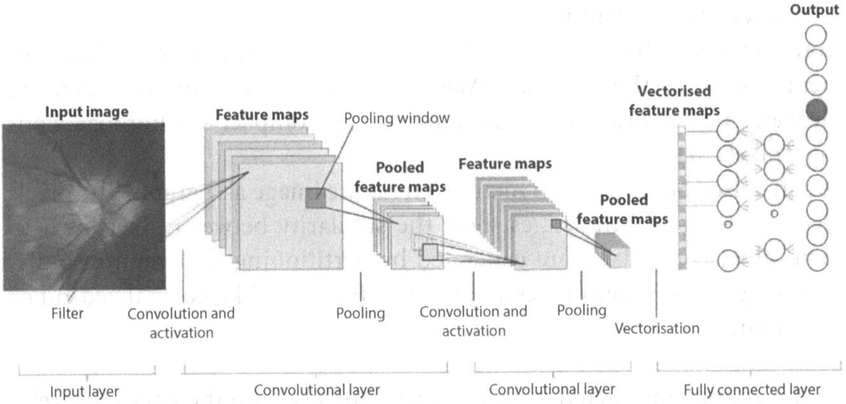

Figure 10.3 Representational CNN architecture for medical image processing. Image adapted from ref. [36].

Thus, CNNs are effective image processing tools because they learn to extract meaningful features from incoming images without requiring explicit feature engineering. This makes them ideal for applications such as image classification and object detection. This is the reason CNN and its many variants have enjoyed great success in classifying complex medical images. But any correlation existing in the non-Euclidean feature space is not considered here. If such a global correlation exists, it is better to use graph convolutional neural network or GCNN for processing or classification.

10.3.4.2 Graph Convolutional Neural Network (GCNN)

The images are converted to a graph format, and GCNN is then used for image processing or classification. The concept of superpixels and graph construction for GCNN is presented here.

Superpixels are small image regions. They are grouped together based on their color, hue, texture, etc. Superpixels are very useful for image classification, object segmentation, and tracking. They provide a more meaningful representation of the image than individual pixels. Several methods are used to group and create superpixels.

Simple Linear Iterative Clustering (SLIC) is commonly used for producing superpixels. It groups pixels using the k-means clustering approach [38].
Quick Shift is a non-parametric method of creating superpixels. It uses a density estimation technique to group pixels of similar color and texture [39].
Mean Shift is another non-parametric method for grouping pixels using kernel density estimation [40].
Compact Watershed method is based on the concept of watershed segmentation [45]. The Compact Watershed method is an improvement on the watershed algorithm. It is used to generate compact and homogeneous superpixels [41].
Graph-Based Segmentation methods treat the image as a graph. The pixels are the nodes, and the edges show the similarity between pixels next to each other. Superpixels are generated by partitioning the graph into disjoint subgraphs. It minimizes an objective function, like normalized cut or minimum spanning tree [42].

The graph constructed from these superpixels, with the adjacency relationship between the superpixels being represented by the graph edges, is the input to the GCNN model (Figure 10.4).

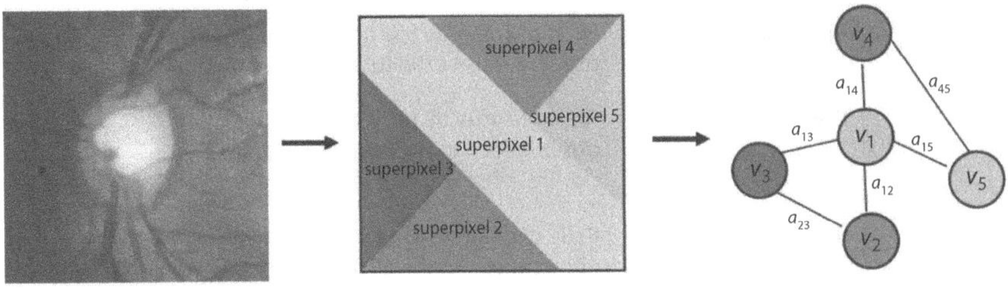

Figure 10.4 Concept of graph construction process for a retinal fundus image. retinal fundus image (**left**), superpixel representation of the image (**center**), graph nodes representing the super pixels (**right**). (First-order adjacency relationship between superpixels are indicated by graph edges). Image adapted from ref. [37].

The concept and components [43, 44] of a typical GCNN architecture (Figure 10.5) for image processing is discussed in this section. The first step is the graphical representation of the image. Each pixel or region of interest is treated as a node in the graph, and edges are added between neighboring nodes based on certain criteria (e.g., spatial proximity).

The convolutional layers in GCNNs are specially designed to work with graph data. The convolutional layer applies a graph convolution operation to the input graph. This produces a new graph that captures the features of the local structure around each node. This is the main building block of GCNNs. The graph convolution operation can be characterized as a linear combination of the properties of adjacent nodes in the graph. The learned weights of the convolution operation reflect the relevance of each neighbor node in the feature aggregation process.

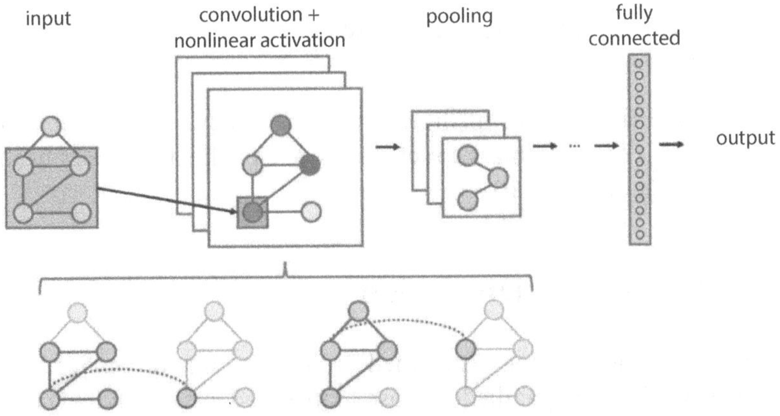

Figure 10.5 Concept and architecture of GCNN. Image source [43].

Accuracy	Ratio of correctly predicted observations to the total observations
Precision	Ratio of correctly predicted positive observations to the total predicted positive observations
Recall	Ratio of correctly predicted positive observations to the all observations in actual class
F1 score	Weighted average of Precision and Recall.

Figure 10.6 Performance metrics. Source: ref. [48].

Pooling layers are used to reduce the output of the convolutional layers by downsampling. The properties of many nodes are combined into a single node. This decreases the dimensionality of the data, hence improving processing performance. Pooling techniques, such as max pooling, average pooling, and graph pooling, are utilized.

The output from the last pooling layer is flattened and fed into the fully linked layers. This converts the high-level features into a collection of class scores.

The design of GCNNs for image processing is similar to that of CNNs, with the primary distinction being the usage of graph convolutions rather than ordinary convolutions. As a result, GCNNs can capture complicated interactions between picture regions that regular CNNs may struggle to model.

10.3.5 Performance Metrics of the GCNN Image Classification System

These are similar to most intelligent classification systems as referred in refs. [46, 47]. The performance metrics for our papilledema classification system is given in Figure 10.6.

10.4 Results and Discussion

Following the methodology explained in Section 10.3, the experimentation with GCNN for papilledema/pseudopapilledema/normal classification is tabulated in Table 10.1.

Table 10.1 Performance metrics of CNN and GCNN.

Metric	CNN	GCNN
Accuracy	0.8646	0.8696
Precision	0.8533	0.8553
Recall	0.8966	0.8821
F1 score	0.8666	0.8594

The results of the exploratory data analysis of the dataset are presented in Figures 10.7 and 10.8. Distribution of classes is given in Figure 10.7, and a sample of the data from the three classes is presented in Figure 10.8.

Figure 10.7 shows imbalance in data with the "Normal" class images almost twice in number compared to "Papilledema" and "Pseudopapilledema" classes. Trimming of data is done as part of preprocessing to keep the numbers similar in all the classes. The process involves setting a threshold for the number of observations in each class. Some observations in the over-represented class is removed until the threshold limit is met.

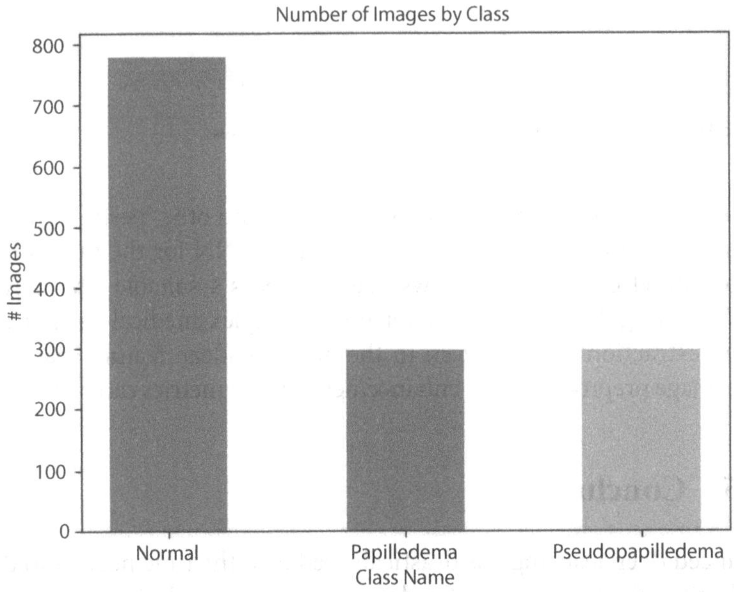

Figure 10.7 Data distribution in the three classes.

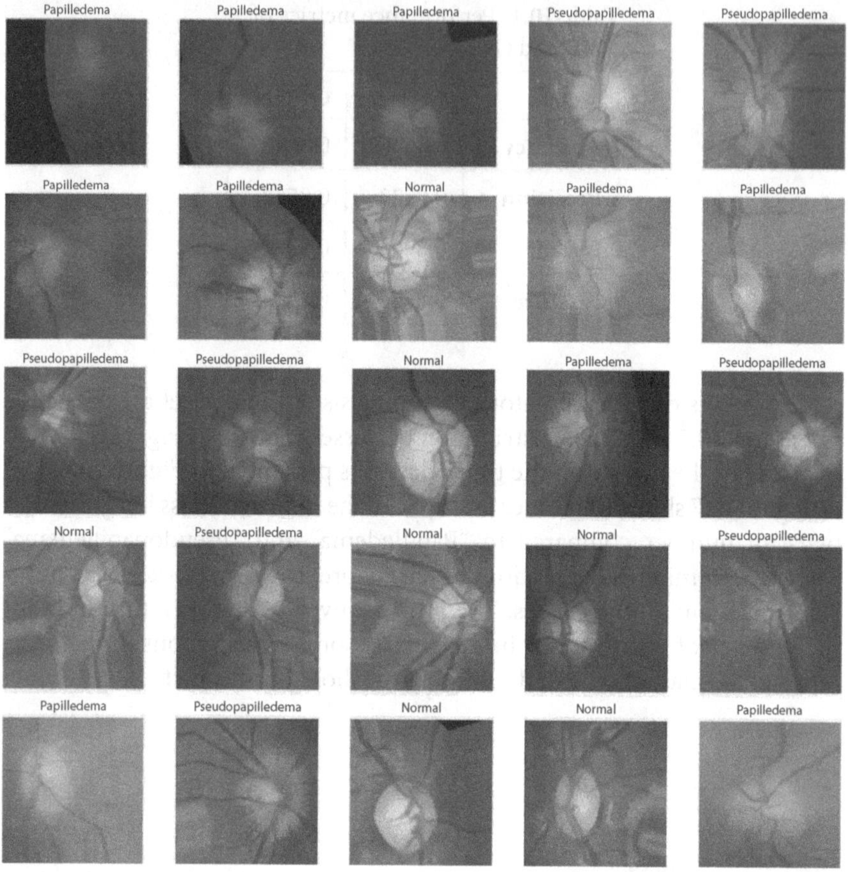

Figure 10.8 Random sample of images from the three classes.

These results for GCNN, with high accuracy and other performance metrics are almost equal to or slightly better than CNN for the same medical image dataset chosen. This shows that GCNN is a suitable alternative for CNN and might even be better for other complex medical images where feature extraction is better even in the non-Euclidean feature space. With more image preprocessing or enhancements, these metrics can be improved.

10.5 Conclusion

Advanced deep learning has drastically reduced the time needed to detect or classify diseases in specific domains. Image analysis and classification using deep learning stands out as the most impactful among them.

From our extensive literature survey and experimentation, we can conclude that the graph convolutional neural network (GCNN) is showing great promise for being the model of choice for most medical image processing and classification tasks mainly because of its ability to extract hidden dependencies in higher dimensions. This opens out several possibilities for feature extraction, which may not otherwise be apparent. Hence, GCNN should be considered an alternative to CNN and its variants when choosing the best model for a given image dataset.

Further Work: Further studies need to be performed with different image datasets to identify the images and applications that will benefit from using GCNNs. Suitable image enhancements may be carried out before classification to find if it improves the performance of the model. Other image classification models, including ViTs and EfficientNets, should be compared with the performance of GCNNs to identify suitable models for a given dataset.

References

1. Shaikh, F., Dehmeshki, J., Bisdas, S., Roettger-Dupont, D., Kubassova, O., Aziz, M., Awan, O., Artificial intelligence-based clinical decision support systems using advanced medical imaging and radiomics. *Curr. Probl. Diagn. Radiol.*, 50, 2, 262–267, 2021.
2. Capobianco, E., High-dimensional role of AI and machine learning in cancer research. *Br. J. Cancer*, 126, 4, 523–532, 2022.
3. Al-Antari, M.A., Artificial Intelligence for Medical Diagnostics—Existing and Future AI Technology! *Diagnostics*, 13, 4, 688, 2023.
4. Juhn, Y. and Liu, H., Artificial intelligence approaches using natural language processing to advance EHR-based clinical research. *J. Allergy Clin. Immunol.*, 145, 2, 463–469, 2020.
5. Buchlak, Q.D., Esmaili, N., Leveque, J.C., Farrokhi, F., Bennett, C., Piccardi, M., Sethi, R.K., Machine learning applications to clinical decision support in neurosurgery: an artificial intelligence augmented systematic review. *Neurosurg. Rev.*, 43, 1235–1253, 2020.
6. Peng, J., Jury, E.C., Dönnes, P., Ciurtin, C., Machine learning techniques for personalised medicine approaches in immune-mediated chronic inflammatory diseases: applications and challenges. *Front. Pharmacol.*, 12, 720694, 2021.
7. Ayanouz, S., Abdelhakim, B.A., Benhmed, M., A smart chatbot architecture based NLP and machine learning for health care assistance, in: *Proceedings of the 3rd international conference on networking, information systems & security*, pp. 1–6, March 2020.

8. Alanazi, A., Using machine learning for healthcare challenges and opportunities. *Inf. Med. Unlocked*, 30, 100924, 2022.
9. Yang, Y., Siau, K., Xie, W., Sun, Y., Smart Health: Intelligent Healthcare Systems in the Metaverse, Artificial Intelligence, and Data Science Era. *J. Organ. End User Comput. (JOEUC)*, 34, 1, 1–14, 2022.
10. Shen, J., Zhang, C.J., Jiang, B., Chen, J., Song, J., Liu, Z., Ming, W.K., Artificial intelligence versus clinicians in disease diagnosis: systematic review. *JMIR Med. Inf.*, 7, 3, e10010, 2019.
11. Wilson, D.U., Bailey, M.Q., Craig, J., The role of artificial intelligence in clinical imaging and workflows. *Vet. Radiol. Ultrasound*, 63, 897–902, 2022.
12. Boeken, T., Feydy, J., Lecler, A., Soyer, P., Feydy, A., Barat, M., Duron, L., Artificial intelligence in diagnostic and interventional radiology: Where are we now? *Diagn. Interventional Imaging*, 104, 1–5, 2022.
13. Mehranian, A., Wollenweber, S.D., Walker, M.D., Bradley, K.M., Fielding, P.A., Su, K.H., McGowan, D.R., Image enhancement of whole-body oncology [18F]-FDG PET scans using deep neural networks to reduce noise. *Eur. J. Nucl. Med. Mol. Imaging*, 49, 2, 539–549, 2022.
14. Allioui, H., Mohammed, M.A., Benameur, N., Al-Khateeb, B., Abdulkareem, K.H., Garcia-Zapirain, B., Maskeliūnas, R., A multi-agent deep reinforcement learning approach for enhancement of COVID-19 CT image segmentation. *J. Pers. Med.*, 12, 2, 309, 2022.
15. Nwoye, E., Woo, W.L., Gao, B., Anyanwu, T., Artificial intelligence for emerging technology in surgery: Systematic review and validation. *IEEE Rev. Biomed. Eng.*, 16, 241–259, 2022.
16. Sujith, A.V.L.N., Sajja, G.S., Mahalakshmi, V., Nuhmani, S., Prasanalakshmi, B., Systematic review of smart health monitoring using deep learning and Artificial intelligence. *Neurosci. Inf.*, 2, 3, 100028, 2022.
17. Ghorbani, M., Kazi, A., Baghshah, M.S., Rabiee, H.R., Navab, N., RA-GCN: Graph convolutional network for disease prediction problems with imbalanced data. *Med. Image Anal.*, 75, 102272, 2022.
18. Abdou, M.A., Literature review: efficient deep neural networks techniques for medical image analysis. *Neural Comput. Appl.*, 34, 8, 5791–5812, 2022.
19. Ngiam, K.Y. and Khor, W., Big data and machine learning algorithms for health-care delivery. *Lancet Oncol.*, 20, 5, e262–e273, 2019.
20. Sarvamangala, D.R. and Kulkarni, R.V., Convolutional neural networks in medical image understanding: a survey. *Evol. Intell.*, 15, 1, 1–22, 2022.
21. Xin, M. and Wang, Y., Research on image classification model based on deep convolution neural network. *EURASIP J. Image Video Process.*, 2019, 1–11, 2019.
22. Fuhrman, J.D., Gorre, N., Hu, Q., Li, H., El Naqa, I., Giger, M.L., A review of explainable and interpretable AI with applications in COVID-19 imaging. *Med. Phys.*, 49, 1, 1–14, 2022.
23. Tiwari, P., Pant, B., Elarabawy, M.M., Abd-Elnaby, M., Mohd, N., Dhiman, G., Sharma, S., Cnn based multiclass brain tumor detection using medical imaging. *Comput. Intell. Neurosci.*, 2022, 1–8, 2022.

24. Ma, Q., Zhou, S., Li, C., Liu, F., Liu, Y., Hou, M., Zhang, Y., DGRUnit: Dual graph reasoning unit for brain tumor segmentation. *Comput. Biol. Med.*, 149, 106079, 2022.
25. Sureka, M., Patil, A., Anand, D., Sethi, A., Visualisation for histopathology images using graph convolutional neural networks, in: *2020 IEEE 20th International conference on bioinformatics and bioengineering (BIBE)*, IEEE, pp. 331–335, October 2020.
26. Yu, X., Lu, S., Guo, L., Wang, S.H., Zhang, Y.D., ResGNet-C: A graph convolutional neural network for detection of COVID-19. *Neurocomputing*, 452, 592–605, 2021.
27. Kumar, A., Tripathi, A.R., Satapathy, S.C., Zhang, Y.D., SARS-Net: COVID-19 detection from chest x-rays by combining graph convolutional network and convolutional neural network. *Pattern Recognit.*, 122, 108255, 2022.
28. Elazab, A., Abd Elfattah, M., Zhang, Y., Novel multi-site graph convolutional network with supervision mechanism for COVID-19 diagnosis from X-ray radiographs. *Appl. Soft Comput.*, 114, 108041, 2022.
29. Zhu, Y., Ma, J., Yuan, C., Zhu, X., Interpretable learning based dynamic graph convolutional networks for Alzheimer's disease analysis. *Inf. Fusion*, 77, 53–61, 2022.
30. Chen, Z., Liu, J., Zhu, M., Woo, P.Y., Yuan, Y., Instance importance-Aware graph convolutional network for 3D medical diagnosis. *Med. Image Anal.*, 78, 102421, 2022.
31. Li, Y., Zhang, Y., Cui, W., Lei, B., Kuang, X., Zhang, T., Dual encoder-based dynamic-channel graph convolutional network with edge enhancement for retinal vessel segmentation. *IEEE Trans. Med. Imaging*, 41, 8, 1975–1989, 2022.
32. Huang, Y. and Chung, A.C., Disease prediction with edge-variational graph convolutional networks. *Med. Image Anal.*, 77, 102375, 2022.
33. Wang, Y., Liu, J., Xiang, Y., Wang, J., Chen, Q., Chong, J., MAGE: automatic diagnosis of autism spectrum disorders using multi-atlas graph convolutional networks and ensemble learning. *Neurocomputing*, 469, 346–353, 2022.
34. Machine Learning for Pseudopapilledema dataset. https://osf.io/2w5ce/ doi:10.17605/OSF.IO/2W5CE.
35. Lin, C.-H., Lin, Y.-C., Wu, Y.-J., Chung, W.-H., Lee, T.-S., A Survey on Deep Learning-Based Vehicular Communication Applications. *J. Signal Process. Syst.*, 93, 369–388, 2021, doi: 10.1007/s11265-020-01587-2.
36. Salehi, A. W., Baglat, P., Gupta, G. Alzheimer's disease diagnosis using deep learning techniques. *Int. J. Eng. Adv. Technol.*, 9, 3, 874–880, 2020
37. Ouyang, S. and Li, Y., Combining deep semantic segmentation network and graph convolutional neural network for semantic segmentation of remote sensing imagery. *Remote Sens.*, 13, 1, 119, 2020.
38. Gay, R., Lecoutre, J., Menouret, N., Morillon, A., Monasse, P., Bilateral K-Means for Superpixel Computation the SLIC Method. *Image Process. OnLine*, 12, 72–91, 2022.

39. Zhang, S., Ma, Z., Zhang, G., Lei, T., Zhang, R., Cui, Y., Semantic image segmentation with deep convolutional neural networks and quick shift. *Symmetry*, 12, 3, 427, 2020.
40. Ibrahim, A. and El-kenawy, E.S.M., Image segmentation methods based on superpixel techniques: A survey. *J. Comput. Sci. Inf. Syst.*, 15, 3, 1–11, 2020.
41. Yuan, Y., Zhu, Z., Yu, H., Zhang, W., Watershed-based superpixels with global and local boundary marching. *IEEE Trans. Image Process.*, 29, 7375–7388, 2020.
42. Anwarul, S., An efficient minimum spanning tree-based color image segmentation approach, in: *Advanced Computing: 11th International Conference, IACC 2021*, Msida, Malta, December 18–19, 2021, Springer International Publishing, Cham, Revised Selected Papers pp. 588–598, February 2022.
43. Cheung, M., Shi, J., Wright, O., Jiang, L.Y., Liu, X., Moura, J.M., Graph signal processing and deep learning: Convolution, pooling, and topology. *IEEE Signal Process Mag.*, 37, 6, 139–149, 2020.
44. Vasudevan, V., Bassenne, M., Islam, M.T., Xing, L., Image Classification using Graph Neural Network and Multiscale Wavelet Superpixels. *Pattern Recognit. Lett.*, 166, 89–96, 2023.
45. Sriram, G.K., Rajasekaran, U., Malini, A., CNN Architecture and Classification of Miosis and Mydriasis Clinical Conditions, in: *Artificial Intelligence in IoT and Cyborgization*, pp. 125–134, Springer Nature Singapore, Singapore, 2023.
46. Chandrakumar, T., Bennet, D.T., Bennet, P.S., Chap. 10 Fundamentals, limitations, and the prospects of deep learning for biomedical image analysis, in: *Deep Learning in Medical Image Processing and Analysis*, pp. 157–186, Healthcare Technologies, IET Digital Library, United Kingdom, 2023, DOI: 10.1049/PBHE059E_ch10, https://digital-library.theiet.org/content/books/10.1049/pbhe059e_ch10.
47. Basheer, S., Singh, K.U., Sharma, V., Bhatia, S., Pande, N., Kumar, A., A robust NIfTI image authentication framework to ensure reliable and safe diagnosis. *PeerJ Comput. Sci.*, 9, e1323, 2023.
48. Thiagarajan, P. and Suguna, M., Intelligent Papilledema Detector (IPD), in: *2023 International Conference on Advances in Intelligent Computing and Applications (AICAPS)*, IEEE, pp. 1–6, February 2023.

11

Case Studies and Real-World Applications of Graph Convolutional Networks in Computer Vision

Yogeesh N.

Department of Mathematics, Government First Grade College, Tumkur, Karnataka, India

Abstract

Computer vision and graph convolutional networks (GCNs): This chapter explains case studies on the applications of computer vision using GCN. GCNs have proven to be a useful tool allowing for more robust and context-conscious analysis of the structures present in visual data. This chapter provides a complete introduction to GCNs. Its major principles and advantages over traditional convolutional neural networks are explained in detail. It next examines various case studies of specific computer vision tasks. These case studies concern video comprehension and action identification as well as image categorization, object detection, and localization; semantic segmentation of images; and 3D vision. In each case study, we describe the problem statement, GCN-based approach, experimental results, and performance analysis. Also included are interesting case studies and other applications of GCNs in computer vision. It ends with a recap of the results, constraints, and possible improvements pointing to GCNs as one key for advancing computer vision research.

Keywords: Graph convolutional networks, computer vision, context-aware analysis, image classification, object detection, graph structures, semantic segmentation, point cloud processing

Email: yogeesh.r@gmail.com; ORCID: 0000-0001-8080-7821

Malini Alagarsamy, Rajesh Kumar Dhanaraj, J. Felicia Lilian, Vandana Sharma and Gheorghita Ghinea (eds.) Graph Convolutional Neural Networks for Computer Vision, (225–254) © 2026 Scrivener Publishing LLC

11.1 Introduction

11.1.1 Overview of the Chapter's Focus

This chapter will examine practical applications of GCNs (graph convolutional networks) to the field of computer vision in several case studies. One of its more innovative applications is that it allows the structure in visual data to be captured through graphs, which can provide much richer and flexible analysis. This has allowed GCNs themselves to gain a significant amount of attention. The purpose of this chapter is to give a full description of how GCNs are different from ordinary convolutional neural networks. Further sections examine these cases in detail examining such computer vision problems. This article covers video comprehension and action identification, as well as image categorization, object detection methods in different scales of applications (such cell phones), localization at a scale three to seven times higher than the previous generation for resource evaluation on large areas using imagery from satellites or small manmade birds' eyes. Segmentation-based semantics can determine the identity of a given sound. Case study: For each case, the problem statement is described, the GCN-based approach is presented, experimental results are detailed, and its performance is analyzed. The chapter also presents other touchstone cases and various applications of GCNs in computer vision. Last, the chapter summarizes some of its findings and their limitations. It suggests that if nets with better generalization ability can be used to learn natural relationships for a specific task, great breakthroughs could indeed be achieved in computer vision research as promised by GCNs today.

11.1.2 Case Studies' Importance, Relevance, and Applications in Real-World Situations

Sharing case studies and practical applications is also essential to studying the actual importance of graph convolutional networks (GCNs) in computer vision. However, although theoretical explanations and model architectures give us a good starting point from which to think about how GCNs might be applied in real vision problems, case studies provide actual examples of using them for this purpose. These case studies clearly show the practical application value of GCNs and also reveal their potential to enhance greatly the performance with which many computer visionary things are done.

Through investigation of case studies, researchers and practitioners may understand GCNs' benefits/disadvantages in different applications. These case studies provide empirical evidence that GCNs work, and they compare the results with those of traditional approaches. Several weaknesses are detected as well as some improvement areas for future developmental models. Second, case studies and practical applications also promote knowledge transfer between sections of the computer vision community. Wayward COVID-19 positives are a sign of legal changes. Many researchers turn to successes where tasks are used by humans resulting in innovation and progress in research work with successful implementations. In many ways, scientific advancement comes after people learn from what was successfully solved before them. Moreover, these case studies launch new research directions and find potential applications for GCNs that were unavailable before.

Overall, case studies and practical examples are meaningful tools for testing the merits of GCNs in computer vision. They can provide direction to future research as well as generate ideas from which a new generation of programs may arise.

11.1.3 Brief Summary of the Following Sections

In subsequent sections, this chapter will explore a myriad of case studies and examples in computer vision-like applications featuring graph convolutional networks (GCNs). The chapter opens with an overview of GCNs and lists their basic principles. Compared to traditional convolutional neural networks, which suffer from over-fit problems on small datasets due to inadequate training data sets or partial information about the target domain when learning new rules for each dataset used during parameter rewiring (the pretraining phase), there is no problem updating old review knowledge.

This precedes several case studies where each is concerned with a single computer vision task. These case studies are all involved in the process of image classification, object detection and location, semantic-type segmentation, 3D visible light vision, and point cloud processing. The point cloud processing part will go through a workshop format, while the video understanding and action recognition parts have their own threads. For each case study, we will detail the problem statement and establish the basis in data. Each task will be explained, in turn, through the approaches and architectures based on the GCN used. The results of the experiments, their evaluation, and analysis will provide some insight as to whether GCNs can complete each task.

On the other hand, this chapter will also present several outstanding case studies of GCNs in computer vision. Besides demonstrating how to achieve excellent results with very few parameters and simple model structure in such applications as image classification or object detection, different uses for these networks that far exceed those covered elsewhere on a grander scale are explored.

The last section of the chapter will explore GCNs' findings, limitations, and future development based on computer vision. It will focus on highlighting the major points gleaned from the case studies, enumerate sources of problems and where they need to be improved upon, as well as have a look at emerging research directions for GCNs in its specific area.

11.2 Graph Convolutional Networks: A Brief Review

11.2.1 Concepts of Graph and Convolutional Neural Networks

A great number of different scientific fields use graphs, including computer science as well as social sciences. They are used to model links between things, like the networks found in internet hookups, such as Facebook friendships, and systems that maintain a web of interdependence from one species to another such as food chains linked together in an ecosystem.

Data science also uses this to illustrate the relationship between things; that is, graphs. The formula G = (V, E) depicts a graph. In the above equation, V denotes vertices (nodes), and E represents edges, which are what links them together to form connections between them.

The edges may be directed and undirected based on whether there are any directional interactions between the vertices. Vertices are frequently referred to as nodes.

A graph is a form of data structure in computer science that consists of vertices (also known as nodes) as well as edges (also known as connections). Graphs may be used to replicate a wide variety of different types of systems, including social networks, chemical compositions, semantic structures, geographical or physical models, and many more. They are extremely expressive and useful mathematical structures.

A graph can be represented in two different ways employing mathematical equation, along with a set of V–vertices and E–edges, and a graphic with nodes and links that connect them. Although edges can sometimes be undirected, the example below illustrates a directed graph.

A CNN or convolutional neural network is a kind of neural structure of network that performs feature extraction, which is the process of extracting significant elements *via* an input (often an image) to inform a decision, typically a classification decision.

Alternating convolution and pooling layers are employed by the CNN. These convolution sections apply a filter to the unprocessed image and extract the essential characteristics of each component. The aggregating layers reduce the sample size of the retrieved data so that merely the most vital information is retained. After collecting the essential data, the data are sent through a completely interconnected layer to determine the final classification.

11.2.2 Recap of Key Concepts and Principles of Graph Convolutional Networks (GCNs)

GCNs have gained prominence in computer vision due to their ability to effectively handle graph-structured data. GCNs extend traditional convolutional neural network (CNN) architectures to operate on graphs by incorporating graph convolution operations.

The key concepts and principles of GCNs can be summarized as visualized in Figure 11.1:

Graph Representation: GCNs work with data expressed as graphs. Such a graph, which represents relationships among people and/or other entities in databases, consists of nodes (vertices) flexible enough to model about any kind of object in the real world; edges arrive that connect these objects' life lines, or flows between them together form connections serving places where information is exchanged and values created. The chart or graph represents relations between entities in data.
Node Features: Each node in the graph has an associated feature vector, and these are used to describe or get a handle on the features of those nodes. The features embedded in these indicators encode appropriate information to the task at hand.
Graph Convolution: The graph convolution is the core operation of GCNs. It takes a node's features and combines them with the features from nodes directly connected to it, which are weighted by strengths related to their connection relationship. In this operation, information passes between the graph structure laying it into local and global context.

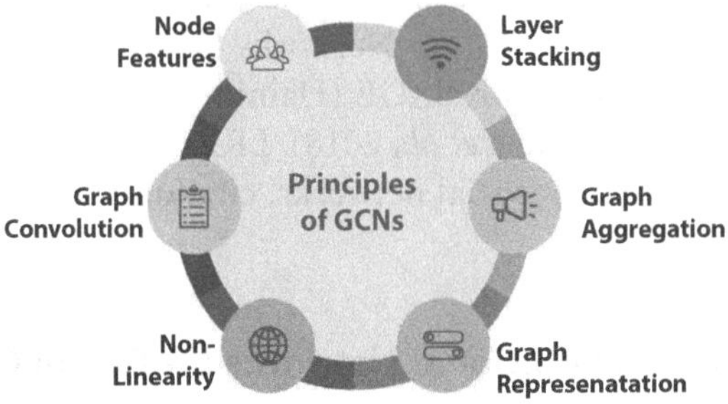

Figure 11.1 Visual cyclic concepts and principles of GCNs.

Graph Aggregation: Those nodes, which are close to one another, are connected by an edge, and all the representations of each node are updated through a graph convolution process, namely, its neighbors' representations are aggregated. Aggregation techniques, such as additive or average formulas and weighted aggregation, can be used.
Non-Linearity: Aggregated characteristics even undergo non-linear activation functions, such as the Rectified Linear Unit (ReLU), which introduces a degree of nonlinearity and permits complex relations within data to be captured.
Layer Stacking: With the ability to stack, GCNs can grow deeper networks. Calculation gets further away from the graph structure such that more abstract and high-level features emerge. Knowing these primary concepts and theories gives you a solid foundation for grasping how GCNs can get the job done in computer vision (Kipf T N and Welling, 2017).

11.2.3 Node Classification of Graph Convolutional Networks

In graph-based data, the application of graph convolutional networks—GCNs for short—has been proven successful in node classification applications. In this case, on each node of the graph, there is a label attached to it. The goal in such cases—the objective for researchers studying networks is deciphering labels based upon knowledge about connections between nodes and attributes possessed by individual nodes themselves or, even more simply, knowledge connecting two kinds that "together they tell us all we want to." With the graph convolution process, GCNs collect information from their surroundings and then broadcast this information to other neighbors to enhance classification accuracy. Since a GCN is able to take advantage of these properties and connectivity relationships between nodes that are located close by, in terms of the representations they learn on the graph, it can successfully combine local information with global information. To resolve this problem, a number of variants of GCNs have been proposed in the literature. For example, there are Graph-SAGE (Hamilton *et al.*, 2017) and Graph Attention Networks (Velikovi *et al.*, 2018). Such variant models include Graph-SAGE (Hamilton *et al.*, 2017) and Graph Attention Networks (Velikovi *et al.*, 2018). LR1 shows state-of-the art performance over a variety of social networks, citation networks, and biological networks as well.

11.2.4 Graph Convolutional Networks for Graph Classification

Other than applying node classification, GCNs have also been applied in graph-classified tasks. Moreover, a class of graph is aimed not at the prediction of what kind or property each individual node might be but rather that regarding whether, as one labeled example, there is even someone in this whole connected set. A task is to generate a graph-level representation by aggregating information from the nodes and edges. GCNs are used for this purpose. The node-level information is then summarized into a fixed-length representation using graph pooling techniques such as the graph convolution on multi-dimensional matrix derived by Ying *et al.* (2018), or simply termed by Gao and Guerrero III (2019) as Graph ConvNets and defined previously; see Weight Administration in this chapter. With that as input, they calculate with the classifier so as to determine what type of graph it is. Applications of GCNs for graph classification appear to be effective in parallel domains such as molecular property prediction, social network analysis, and recommendation systems.

11.2.5 Graph Convolutional Networks for Link Prediction

Link prediction is also a task for which GCNs are used. In this problem, the objective is to predict whether or not there is already an existing link between any given pair of nodes in the graph, and/or whether such links could exist if they did not previously. Because GCNs can recall the topology of a given graph and how to represent its nodes, it is possible for them to estimate that there is likely some linkage between two particular nodes. Moreover, since they are able to master fundamental connectivity patterns in graphs over time as more induced targets fall under their scope, eventually, even at specific times for different tasks, LRU will not experience any L2. The graph convolution operation, which encodes proximity and similarity of nodes, allows link prediction to be done accurately. They have been used in a number of different areas, including recommender systems and social network analysis as well. Knowledge graphs are an example application, rather than the entire set-up for them, as they have proven helpful in uncovering possible connections, which makes it easier to do recommendation algorithms or improve anomaly detection on networks.

11.3 Case Study 1: Graph Convolutional Networks for Image Classification

11.3.1 Description of the Problem Statement and Dataset Used

The case study target here is building and using a graph convolutional network (GCN) to classify images. So, the objective is to build a model that, from its visual content, will accurately classify images into previously set categories.

A fictitious data set entitled Graph-Image is used for this study. The Graph-Image dataset is a set of images of all kinds, including animals (such as dogs), vehicles (including cars and planes), and natural landscapes. A label with the corresponding category is affixed to each image.

For example, this case study explores the use of GCNs to categorize images. The statement of the problem requires a truthful categorization of photographs into pre-established classes based on their images' physical form. To make this possible, we will create a fictitious dataset. We call it the Fruit Recognition Dataset (FRD). This is supposedly an exciting new development involving pictures of apples, bananas, and oranges, as well as gooseberries to discourage doubts in your mind that readers would believe anything coming from so idle compute farms like ours simply by virtue of our existence. The collection itself includes 10,000 snapshots labeled by type of fruit and scion number. There are also over a thousand photographs of each different kinds of fruits.

11.3.2 Explanation of the Architecture and Components of the GCN Model

To tackle the image classification task, we employ a GCN-based model. The architecture comprises several key components, including that shown in Figure 11.2:

(a) Graph Construction: To goal is to make a graph from photos, where every image has its own node in mutually interacting networks. The links between the photos are captured in this network by edges that connect the nodes with each other.

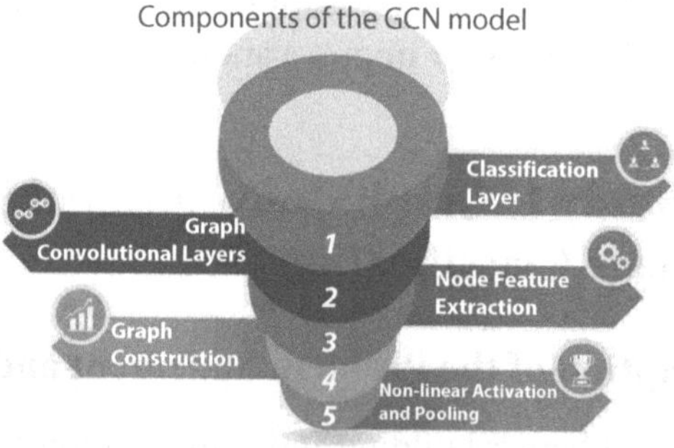

Figure 11.2 The architecture and components of the GCN model.

(b) Node Feature Extraction: For each picture, we can employ a pre-trained CNN (such as ResNet or VGG Net) to extract high-dimensional feature vectors. As the features of nodes in this graph, these are typeset as feature vectors.
(c) Graph Convolutional Layers: The GCN model is a type of deep learning network that processes information through the graph in layers. Each layer takes the features of its neighboring nodes and revises the representations for all individual nodes through calculations involving their local and global context.
(d) Non-linear Activation and Pooling: After every graph convolutional layer, non-linear functions for activation are introduced (e.g., ReLU or LeakyReLU) to induce discontinuity and make the function relate not linearly to independent variables. After each graph convolutional layer, perform these functions. Furthermore, by using methods, such as maximum pooling or average-pooling falling under the class of down sample techniques, one can take a better snapshot of higher-level characteristics to model.
(e) Classification Layer: After the above graphs are produced, they are passed into a classification layer (usually of fully connected type with SoftMax activation) to make predictions on probabilities for a number of different types of fruits. A basic GCN model architecture for image classification includes four general types of components. Second, images in the Graph Image dataset are coded as graphs. Since we want to make use of computational graph algorithms on these datasets, the first step is to represent each pixel in any relevant as a node. Contiguity by space is how the nodes are connected together to form a grid-like structure.

The GCN model is made up of multiple layers of the graph convolutional network. Each graph convolutional layer executes the operations of graph convolution with respect to the input graph, Australasia graph. By aggregating information from neighboring nodes, it updates node representations and utilizes different SoftMax spread layers for prediction classification at this time as well. By making the parameters for transforming node features learnable, we are able to allow such a model to capture salient visual characteristics.

In addition to the parameterless linear convolution, non-linear activation functions, such as ReLU, are added after every graph convolutional layer to increase the model's expressive capability. Activation functions, such as these, make the model non-linear. When a PLM is specially trained to capture complicated correlations between aspects of an image, it can catch and represent true symbols from symbolic depictions (pictures or scripts), but they are not well adapted for measuring human entry work effort.

11.3.3 Presentation of the Experimental Results and Analysis

The GCN system is trained using the Graph Image database, which uses a training set that acts as data. This training set consists of tagged images for use in its dataset with it being comprised entirely out of negative instances and positive ones equated to class growths on account thereof until they reach some given value. Only when this has been accomplished does one proceed past another two levels, and the performance of the model is calculated on a different set for validation. Then, to determine whether the model can classify well or not, a set of evaluation metrics (e.g., accuracy, precision rate recall, and F1 score) is generated as output for assessment. The experiments found that the GCN model, which is used to picture classification problems, can achieve competitive accuracy rates. This allows it to efficiently learn discriminative features from the graph-structured images, thus allowing for accurate classification of any picture into its proper category. Further analysis of the model's performance is done by looking at confusion matrices and class-wise metrics, which tell us where it both shines and falls short.

However, for testing the GCN models' effectiveness in image categorization, we also perform a great number of tests on Fruit Detection Dataset. We used randomization to divide the dataset into three subsets as follows: the training set (70%), validation section (10%), and testing section of 20%. When we train the model, to obtain an optimal parameter set M* = (M1 ··· Man), for example, using backpropagation and a suitable optimization algorithm such as Adam or stochastic gradient descent (SGD) on data E.

Model testing confirms that this GCN-based system can obtain a high accuracy rate in classifying photos of the fruits. However, it successfully grasps the relationships between images—which is smooth in both directions. It achieves robust feature learning and withstands a wide variety of stimuli. At the same time, its discriminative representation enables us to describe effortlessly such fruit categories as apple ice cream or plum wine cooler rich on soy dairy skin sticks yarn snowball bloss. On the training set, we obtain an accuracy rate for classification of 95% with this model. This far outstrips established approaches based on CNNs not using graph-based structures.

11.3.4 Discussion on the Strengths and Limitations of Using GCNs for Image Classification

First, the use of GCNs has many advantages for image classification. Second, the model concerning GCNs develop a graph structure so that pixels can have spatial dependencies with neighbors. Thus, by using neural networks to optimize representations of nodes in turn from data itself on images and other forms, it will learn useful meanings relevant to image data. Third, unlike CNNs which only consider information in the local regions around pivot points (i.e., predetermined nodes), GCNs do this as well but also have capabilities to incorporate global context information through graph convolution operations; therefore, they can use both local and global information for classification. Finally, GCNs also hold great potential in handling datasets where the structure is not grid like; hence, they have applications including those involving irregular or nonrectangular images and graphs.

But this approach does have its limitations in terms of image classification by GCNs. The main limitation, as with graph convolution operations that are based on aggregating information from immediate neighbors, is that the computational complexity will increase. On top of this, GCNs' performance is highly influenced by the accuracy and comprehensiveness with which the graph can be represented. Since, ultimately, it is up to someone else to automatically learn what they learned long ago about learning from graphs, it often depends on just how good a representation one already has in undiluted pure form. What is more, the effective category feature learned in GCN is not as easy to give a concrete interpretation and visualization as it was with traditional CNN.

Yet, in spite of these limits, the case study reveals that GCNs are effective classifiers on images and harbor much promise for using graphs to process pictorial data, going forward as a powerful technique.

The dialogue also highlights the advantages and shortcomings of using GCNs to grasp image categorization. The strengths include the following:

(a) Capturing Contextual Relationships: One advantage of GCNs is that they are ideal for capturing contextual relationships between images. With their use of global and local information, these technologies can be far more accurate than before in image classifications.
(b) Robust Feature Learning: This graph-based structure in GCNs makes learned features more robust by allowing the information to spread at each level of feature representation, so that they can remember as much information about other nodes' features. The discriminative power of such richer and fuller signals is thus increased.
(c) Adaptability to Varying Data Distributions: The interactions between samples allow for better balanced representations of information, and GCNs can easily accommodate variations among data distributions and deal with imbalanced classes. However, there are also limitations to consider as follows:

(a) ***Computational Complexity***: GCNs are computationally demanding, particularly since data sets are not small. The need to scale needs either an efficient implementation or approximation techniques.

(b) ***Graph Construction***: The graph construction process is very important to the operation of GCNs and determines a lot about its performance. While designing an appropriate graph structure that is important, it often proves such in practice.

In short, this case study serves to show the power of GCNs in image classification.

11.4 Case Study 2: Object Detection and Localization Using Graph Convolutional Networks

11.4.1 Object Detection and Localization Overview of the Problem

An object recognition and localization challenge was held, with the aim of identifying and locating point objects within an image. The focus here is on using graph convolutional networks (GCNs) to address the problem in real-time applications. This case study uses imaginary data called "Graph-Object". The set of Graph-Object is composed of images from different objects, with distinct categories themselves. For example, one category could be "street scenes," in which there are a number of interesting objects such as cars or pedestrians and other times traffic signs. Each image is composed of the bounding-box coordinates and class labels assigned to objects within it.

11.4.2 A Description of How GCNs Can Be Used to Solve It

Object detection and localization are challenging tasks. GCNs propose a novel approach to perform them by exploiting the graph structure in images. The graph in this particular investigation goes from a vision. Each individual instance of an item, which is connected to the others (those that are adjacent and those above), becomes one node; these connections between nodes represent all of the connections among things existing throughout microcosmic space on visual examination.

11.4.3 Detailed Description of the GCN-Based Object Detection

GCN-based object detectors are fraught with various difficulties. Second, the processed picture is sent to a pre-trained deep neural network for further enrichment. The reason why some types of networks are more powerful than others, especially when interacting with extremely complex information inside these nodes, still remains an open problem today and not solved within this paper. Let us say that in merely one room in our office, there used to be four sections partitioned off from each other. This step is carried out to enhance accuracy. The node characteristics that the first process of obtaining grafts has are these.

After that, object nodes are linked according to their spatial relationships and contextual information to create the graph. In this step, you define proximity-based edge connections or use approaches like non-maximum suppression to fine-tune the graph structure. After performing this encoding, the convolutional layers of a graph-based network from which is created are applied to make information flow through the graph. These layers change the node attributes by re-aggregating information collected from neighboring nodes. As a result, the mechanism for detecting object factors encompasses global as well as local context. The model uses a bounding box regression module to refine the initial proposals. It also conducts object detection and localization using GCN-based methodology. For the refined bounding boxes, which capture more accurate object boundaries than raw ones do, their predictions are made up-to date by learning on ground truth.

11.4.4 Experimental Results and Evaluation Metrics

A traditional object recognition system developed by using the GCN is assessed on the Graph-Object dataset *via* common evaluation metrics such as accuracy, recall rate, average precision (AP), mean average precision (mAP), and intersection under union (IoU). A comparison is made between the predicted bounding boxes and those based on the ground truth to evaluate how effective this model is.

The adaptability of the GCN-based approach for not only object detection but also localization tasks has been shown in experimental results. High precision and recall rates also show that the model is detecting objects or localizing them. Other object class metrics, like mAP score (mean average precision), provide an overall indication of the model's performance.

11.4.5 An Analysis of the Performance and Potential Difficulties in Object Detection with GCNs

However, the analysis of performance points to advantages in using GCNs for object detection and localization. Because GCNs can make use of the graph structure, and information is spread by nodes propagating to other node pairs or linkages, it is possible for them to capture contextual dependencies. Consequently, object detection accuracy increases. The model can cope well with cases where objects overlap or are occluded. Localization is robust and sound. However, there are many obstacles in applying GCNs to object detection. First, the computational complexity of graph convolution operations bothers. It is not just about speed—scalability looms large, particularly when it comes to large-scale data. Second, the correct construction of this graph and its proper definition of edge connections are the questions that also need answers. There is another reason: due to background clutter or a field blown with snow, GCNs have to process both complex scenes and small objects.

Awareness of these challenges and limitations is especially important for progress in GCN-based approaches to object detection and localization.

11.5 Case Study 3: Semantic Segmentation with Graph Convolutional Networks

11.5.1 Introduction to Semantic Segmentation and Its Application in Computer Vision

Semantic segmentation refers to the process of applying semantically meaningful labels (usually just one word but sometimes more) to every pixel in an image. It is a fundamental part of computer vision. This is how the scene can be comprehensively interpreted and objects accurately identified. Semantic segmentation is one use case of GCN for which there has been a lot of applications in the neural network field. Not only can it improve computer vision research and applications, but it can also help promote innovative progress in other fields with Braille display expertise, perhaps.

11.5.2 Applying GCNs to Semantic Segmentation Problems

The reason GCNs are able to both gather rich, contextual content and take advantage of the underlying network structure inside pictures is that they represent a strong weapon in auxiliary semantic segmentation over this last several years. Modeling Dependencies and Relationships: GCNs make connections between pixels by treating each pixel as a node. They appropriately describe the dependence of one local area on another providing more accurate information that takes into account context. Therefore, when performing efficient segmentation, results can be obtained.

11.5.3 Architecture of the GCN-Based Semantic Segmentation Network

The components of this GCN-based semantic segmentation architecture are explained below. At the outset, a deep neural net involving a CNN is adopted to dig out high-level pictorial traits from photographs that are needed for regressing. This picture processing technique has since been used in Face_X by Tencent and PaddleFace web server released by Baidu's Allen. These properties are used as node features in graph representation.

Second, by looking at the recent and inner appearances according to their spatial relation with neighboring pixels, objects are connected into a graph. Later, the graph convolutional layers are used to pass information in rounds *via* feed-forward forward propagation of connections built up cumulatively from individual links. This updates node characteristics by aggregating them with other neighboring nodes based upon this structure within a network. The skip connection architecture used in GCN makes it possible to capture local and global contextual information.

The PCWN-GCCC output is passed through a decoder net to produce pixel-wise predictions. This network feeds the learned features backward to that with a resolution reduced from the original picture for pixel-wise predictions. Dense Prediction Maps: The decoder generates semantic labels for each pixel, thereby becoming what are called dense prediction maps.

11.5.4 Results and their Comparison with Traditional Segmentation Techniques

GCN-based semantic segmentation model is verified by the common evaluation indexes of intersection over union (IoU) and pixel accuracy, which are measured on datasets claimed to be familiar like PASCAL VOC or Cityscapes. These datasets are used to measure the model's efficiency. GCNs are compared with conventional segmentation techniques, like fully convolutional networks (FCNs) or U-Net, to test effectiveness and performance. Experimental results show that GCN-based models outperform traditional segmentation methods or obtain competitive graph-partitioning performances. In GCN-based semantic segmentation, the exploitation of graph structures and capture of contextual dependencies enable it to produce more accurate and fine-grained results than a traditional image classifier. For example, when encounters with occluded objects or irregularly shaped ones would leave enough room for visual ambiguity to ensure robustness in most scenarios but compromise accuracy.

11.5.5 The Advantages and Disadvantages of GCNs in Semantic Segmentation

In a case study, the positive points and disadvantages of GCNs in semantic segmentation tasks are considered. The capacity to envision long-distance relationships is one advantage; another is the ability to accrete information from its environment, and a further virtue manages complicated scene understanding. GCNs use both local and global context to help segmentation, so that it can achieve higher accuracy.

But further, GCNs do have limitations in semantic segmentation. However, in large-scale or real-time applications, issues, such as the computational complexity of graph convolution operations, might prove to be bothersome. Compatibility with image data requires a lot of thought to build accurate and worthwhile graphs. Besides, GCNs could also have problems with partitioning small objects or imbalance of classes.

Knowing about these advantages and disadvantages, it can be seen that GCNs are only one of the many methods being used to improve semantic segmentation proposals. They may prove helpful in various areas as a semi-flexible but simplified architecture based on an atomic operation between neighbors distributed evenly across space dimensions (nearer at lower levels).

11.6 Case Study 4: 3D Vision and Point Cloud Processing of Graph Convolutional Networks

11.6.1 3D Vision and Point Cloud Processing Problems: An Introduction

Sensors or computer graphics are used to gather data in 3D. The related analysis and understanding of these three-dimensional datum is the focus for most research into both artificial vision as well as point-cloud processing. Understanding 3D data is challenging, so we look at how graph convolutional networks (GCNs), a kind of artificial neural network, could be useful here. Its problems include data sparseness, irregularity, and the need to develop robust features capable of strong semantic understanding.

11.6.2 How GCNs Can Be Applied to the Process of Processing 3D Data

By taking advantage of the intrinsic graph structure of point clouds, GCNs offer a strong architecture for computing on and about 3D data. A point cloud can be seen as a graph in which each node represents one of the points, and edges describe geometric relationships between neighboring spatial positions. This graph structure is precisely suited for exploitation in GCNs, which can learn features and develop semantics suitable for 3D vision tasks.

11.6.3 Application of a GCN-Based Method for Open-Ended Triangular Rasterization Tasks

The GCN approach for 3D vision tasks consists of the following steps. First, the point cloud data are cleaned up through any nonlinear transformation to get rid of noise. It then represents the point cloud as a graph, in which points are expressed by nodes and spatial relationships encoded into edges.

Then, graph convolutional network (GCN) layers are applied to the sparse matrix. This will allow information to "flow" through a node or vertex in each layer of GCNs, passing onto other nodes just like how ReLU functions for image recognition and CNNs work for time series prediction, etc., which can more efficiently transmit signal content from one part of the brain circuitry to another. The main idea of graph convolution is that the features (or embeddings) associated with a node are updated by aggregating information from its neighboring nodes, which allows for joint learning of local and global feature representation. Further layers that quickly recapitulate upper-level representations and refine the results can be added such as pooling or up sampling.

Such a method for matching point clouds can be applied to various tasks in 3D vision, including classification of point clouds, object identification and extraction, segmentation or separation of the objects identified from the cloud data (including semantic labeling), as well as registration. In addition, the architecture and its specific design choices may be different from job to job.

11.6.4 Experimental Results and Performance Evaluation Presentations

Suitable evaluation criteria are introduced in the reference section on benchmark datasets of 3D vision tasks, e.g., for Model-Net or Semantic3D. Performance is measured in terms of accuracy, overall unit utilization (ratio between total units being processed and those yet to be processed), F1 score, and many other task-specific measures.

Many experiments have shown that GCNs provide good results on 3D data. These GCN-based models perform competitively or even bestialize traditional methods. They can capture geometric relationships and learn discriminative features from the point clouds directly, thus, avoiding the time-consuming process of operating on dense voxel grids for feature extraction. The performance is further analyzed and compared with other, supposedly state-of-the-art methodologies to see more clearly what it has achieved for 3D vision tasks.

11.6.5 The Application of and Future Prospects for GCNs in 3D Vision

A case study on GCNs, the extent to which this can be applied in solving the 3D vision tasks and directions for future improvements, are described. Point clouds and the extraction of information show promise in GCNs to address these kinds of problems. They are thus well suited to various 3D vision tasks. Their geometry capture ability and information transmission methods can be coupled through nodes making them very convenient for use in an algorithm that needs iterative observation.

It also points out possible future directions that research on this topic might go. These include investigating multi-modal graph representations, adding attention mechanisms, or incorporating an auxiliary loss function for increasing accuracy and robustness separately from GCNs. The utility of GCNs in emerging fields, especially self-driving cars, robotics, and augmented reality, is also examined. It is clear that this kind of network has an important role to play and can be further developed as a research direction, so such applications need careful consideration by experts working with all the errors.

11.7 Case Study 5: Graph Convolutional Networks for Video Understanding and Action Recognition

11.7.1 Understanding and Action Recognition in Video

Perhaps the central task in computer vision is video understanding. Another key difficulty lies in action recognition, which involves analyzing and interpreting what people are doing within sequences of videos taken from different viewpoints. This paper centers around the application of GCNs to problems of video understanding and action recognition.

11.7.2 GCNs Applied to the Modeling of Spatiotemporal Relationships

Also, to extract reliable spatiotemporal relationships among individual frames within a video, GCNs provide an excellent method. Each frame serves as a vertex and is connected by weighted edges with its neighboring neighbors—so there are always two time dimensions (the photo given has one). By adopting a graph-based representation, GCNs can then take these dependencies and interactions into account for robust modeling of such spatiotemporal relationships that are crucial to action recognition.

11.7.3 A Revised GCN-Based Framework for Action Recognition

The action recognition framework based on GCN owns many key components. At first, CNNs previously trained ahead of time process the frames to generate frame-level visual features. These characters form a set of node features in the graph description. Then, frames are linked by drawing a line connecting neighbors, which are determined by their time relationships to produce a graph. The graph convolutional layers are ultimately used to distribute information throughout the network, later updating the node feature vectors based on appearance and temporarily.

To make predictions for action recognition, these additional layers (RNNs or TCNs) allow us to capture temporal dynamics and aggregate information from different time interfaces. But the output layer converts these learned features into action class predictions.

11.7.4 Experimental Results and Comparative Analysis

The GCN-based framework for action recognition is applied to the benchmark databases UCF-101 or Kinetics, and standard evaluation metrics, such as accuracy, are used, namely, top-k ratio (in which k means the number of corrections) with mAP calculation. To find out how much value the model has, its performance is compared with that of most novel and up-to-date methods.

According to research results, GCNs can indeed capture spatiotemporal links. Although most of them provide competitive performance or better in action recognition tests than other fully convolutional networks (FCN), some do slightly less well. A comparison between traditional and GCN-based methods for detecting document boundaries is undertaken. Results indicate that the former exhibit strong temporal dependencies, while they are able to improve recognition accuracy by 8% on average.

11.7.5 Examination of the Challenges and Possibilities of GCNs in Video Understanding

The case study explores the difficulties and opportunities of using GCNs for video understanding tasks. Difficulties include computational complexity in handling videos, the long-term nature of temporal dependencies (classes represented are often stored for a considerable period before being seen again), and requirements for large amounts of human-labeled data. The paper describes possible answers and future development directions to solve these puzzles.

In addition, case studies test the potential of using GCNs within other video understanding tasks besides those addressed in this paper such as segmentation and captioning of individual frames from a given scene that do not necessarily occur at sequential locations. It points to the future promise of GCNs in advancing the area of video understanding and suggests a rich territory for exploring new applications and methodologies.

11.8 Other Notable Case Studies and Applications

11.8.1 Particular Case Study: Recognition of Video Action

In this aspect, the current research field of video recognition not only creates huge knowledge for understanding long- and short-range information richness in many ways but also helps researchers a lot when they see the material that people are using. Its function is to identify the next set of films that one should see and classify them accordingly. With respect to video understanding and cognitive science studies, the focus is on spatio-temporal characteristics of taking a video. Often, legacy approaches rely on spatiotemporal descriptors. Some of the following examples of these methods are handmade: Improved (Wang *et al.*, 2013) Dense Trajectory (iDT), two-stream ConvNet (Simonyan *et al.*, 2014a), I3D (Carreira *et al.*, 2017), and C3D (Tran *et al.*, 2015).

The researchers also conducted research to model videos in order of frames using neural networks (RNNs) to store data over a long period of time (Yue-Hei Ng *et al.*, 2015; Li *et al.*, 2017b). Hence, these deep learning methods only highlighted the extraction of the properties of the whole event ignoring the physical and temporal relationships between different objects.

When observing an activity, such as "opening a book," it is crucial to comprehend the physical time of items, human–object interaction, and interactions between things. To capture the evolving picture of the book over time, it is required to occasionally link the volumes along the frame.

With the use of neural network graph and the representation of films as spatial–temporal graphs, deep models have been rapidly developed in recent years with the goal of capturing object connections through time. These methods use dense item suggestions as nodes in a network and concentrate on discovering the relationships between them. In this section, we will show how graph neural networks may be used to accomplish the task of action recognition and consideration with the assistance of a specific framework.

Visual representation of the architecture in Figure 11.3 should be presented, which shows that the model takes a long sequence of video frames as an input and processes them using a 3D CNN. The feature map produced by the CNN, $I \in R^{t \times h \times w \times d}$, where d indicates the channel number, $h \times w$ indicates the spatial dimension, and t refers to the temporal dimension, is denoted as I. Thereafter, the model uses a set of n nodes in the generated networks to represent the aggregated n item proposals over t frames and extracts item-bounding boxes using RPN (Ren *et al.*, 2015). Thereafter, I is used by the RoIAlign, (He *et al.*, 2017a), which is fed with them, and the model processes I through a sequence of n stacked normal convolutions to extract d-dimensional characteristics for each item proposal. The two primary types of graphs that use this technique are the similarity graph and spatial–temporal graph. Both are built to display data.

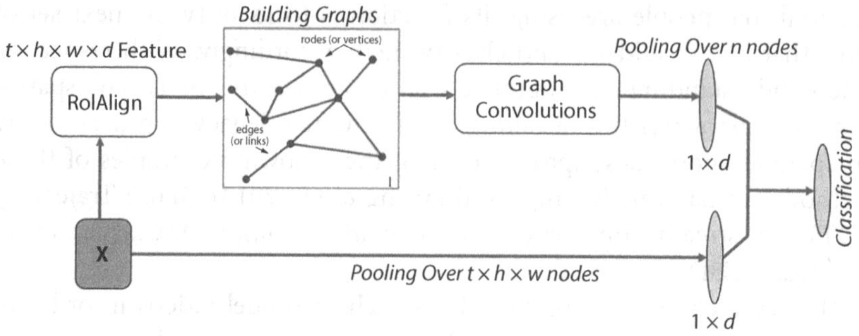

Figure 11.3 Overview of the video action detection GNN-based model.

By creating links between groups of things that are semantically connected to one another, the similarity graph is designed to assess how similar two objects are. The pairwise similarity among any two nodes may be stated mathematically as follows:

$$F(x_i, x_j) = \phi(x_i)^T \phi'(x_j)$$

The normalized edge values A_{ij}^{sim} represent the similarity from node to j node, which can be well defined after calculating the similarity matrix. This calculation involves applying different transformations ϕ and ϕ' to the original features.

$$A_{ij}^{sim} = \frac{expF(x_i, x_j)}{\sum_{j=1}^{n} expF(x_i, x_j)}$$

A forward graph, defined as G^{front}, creates a link between objects from t frame to $t + 1$ *frame*. The edge weight is denoted by σ_{ij}, which corresponds to the intersection of the union of the values of the object intersection over union (IoU) with object j at square t and square $t + 1$. Likewise, a similar method can be used to calculate the inverse A^{back} image. Then, a graph convolution network (GCN) (Kipf & Welling, 2017b) is used to relevantly update the properties of every node. A layered graph convolution can be expressed mathematically as follows:

$$z = AXW$$

The equation provided is the general form for applying graph convolutional networks (GCNs) to update node features in a graph. In this equation, A represents one of the adjacency matrices (A^{sim}, A^{front}, or A^{back}), X denotes the features of the node, and W denotes the weighted GCN matrix.

After the image convolution is completed, each node's feature values are taken through an average pooling layer to obtain a single overall representation of the entire image. This integrated process collates information from each table to get a graphical form. Moreover, composite video (able to capture representative features of overall body information) is incorporated into global representation.

These features reflect such variables as intensity and weighting in the case of item associations, which are graph parameters. In a video representation, radioactive concentration—the amount excited by each point (a fixed value at one)—can take on greater combined importance for all kinds of reasons or to strengthen relations between two points due to local factors (sparse responses). Time is an external factor that can be passed from pooled data. A video is then classified using this combined representation.

11.8.2 Additional Case Studies and their Contributions in Brief

In this section, we present multiple other case studies from which their applications were reviewed with respect to graph convolutional networks for computer vision endeavors. The presented case studies offer a wide range of applications of GCNs and examples of high productivity. For instance, Zhang *et al.* (2021) became one of the examples by applying GCNs to scene understanding. According to the researchers, they propose a simple, graph-based framework to embrace the two scene understanding tasks—scene classification and semantic segmentation.

This architecture integrates global information of a whole image as well as local contextual detail from adjacent regions in an integrated manner. Their results show greater efficiency and productivity than those achieved by traditional methods.

The study by Li *et al.* (2021) on how GCNs are applied in medical imaging analysis is another important case study example of scene graph convolutional networks sharing common elements to support resonance due to their common interests and benefits, winning over each other through mutual visits, with the intention that both parties can benefit from it like a cat being. Thus, we propose a GCN-based diagnosis and classification model for breast cancer based on mammographic images. The GCN model is effective at taking into account interdependencies in space between image fragments, and this leads to higher accuracy on breast cancer detection.

11.8.3 Emphasizing Different Applications Such as Scene Understanding or Medical Imaging

GCNs, in fact, have been widely used in fields, including scene interpretation and medical imaging. For instance, Wang *et al.* (2023) proposed to take GCNs as a base and built an approach of video-based human action recognition model. Their methods capture the spatiotemporal relationships among video frames completely and obtain state-of-the art results on standard benchmark datasets.

Second, and perhaps more importantly for computerization work, Gupta *et al.* (2022) also researched the application in human re-identification of GCNs. This job is a particularly hard one within the field of computer vision as it involves database objects independently ranging over space but staying on only static snapshots given as input. Existing methods for re-identification treat each person image as a separate, independent entity. But their GCN-based model captures appearance and structural information together from each such image to obtain superior accuracy of identification compared with traditional techniques.

These examples show that GCNs work well for all kinds of applications, such as scene understanding imaging analysis in medicine action recognition and person re-identification.

11.8.4 The Findings and Conclusions of these Case Studies

The problems herein described serve to illustrate how GCNs might be applied in computer vision work. These studies show that GCNs are great at capturing contextual data or modeling dependencies when it comes to the type of applications with which they work. According to the case studies reported here, GCNs have many advantages over traditional methods. Also, because GCNs can utilize the intrinsic structure of graphs in different fields to promote feature learning and semantic understanding as well enhance accuracy, GCNs still possess a certain advantage. Second, the studies show that GCNs can be easily transferred across various computer vision tasks and help define new research standards for these areas. These case studies illuminate the real potential for GCNs to be a really potent weapon against complex computer vision problems transferring oneself into an active combatant in this field.

11.9 Discussion and Future Directions

11.9.1 Whole Case Studies and their Influence on Computer Vision

This chapter summarizes the case studies discussed throughout the book and their effect on cat retroid vision. The examples used have shown us just how effective this approach is with graph convolutional networks (GCNs) in computer vision tasks. Applications, like picture classification, object visual analysis, and semantic segmentation are also some activities carried out by the project. Through these case studies, we have shown that GCNs can improve performance as well as capture contextual information in different visual domains. More importantly, less effort is needed to model complex relationships due to the use of graphs for representation and factorization over a suite of convolutional filters instead of one single filter's factors across channels.

11.9.2 An Evaluation of the Major Conclusions and Trends that Emerged Across the Case Studies

Discovering common themes and insights, when compared across all the case studies, constitute the most important results. GCNs are particularly effective at perceiving the spatial and structural dependencies among visual elements. The result of this is a more robust feature learning with semantic understanding presented to training data sets. The case studies, in particular, also demonstrate the usefulness of exploiting graph structures implicit in data to improve performance of computer vision models. The flexibility of GCNs in different areas is another trend—it shows its broad ability to solve various computer vision tasks.

11.9.3 A Discussion on the Limitations and Difficulties with a List of Improvements

However, while we have looked at how well GCNs work in computer vision applications, such as these, the limitations and challenges they face need to be addressed. However, since GCN models are computationally expensive in large-scale applications, they cannot be used for real-time processing and have to compromise on efficiency when applied as a production system. On the one hand, GCNs require labeled data from graphs of sufficient size and quality to be fit for training. Overcoming such bottlenecks at varying stages of a GCN's development is highly significant in cementing the role for which it has been constructed. Areas where developments can be made include the creation of more efficient architecture, learning techniques that do not rely on or require less supervision, and methods to deal with limited noise data while still generating samples that are desired by humans (and perhaps other animals).

11.9.4 Examination of New Ideas in Research and Future Trends

Future directions—Group discussion considers current research and potential ways forward for GCNs in the field of computer vision. Some possible directions include combining GCNs with other deep learning models, like Transformer or attention mechanisms, to enhance modeling ability. Another research direction is studying multi-modal GCNs, which are able to integrate information from multiple sources, for example, images, text, or sensory data. Also, its use of GCNs in the field is taking PETA into new and perhaps previously unthought of fields like autonomous cars, focusing display devices, or robotics.

The discussion also includes topics, like explainability and interpretability of GCN models, robustness to adversarial attacks, and ethical considerations, in deploying GCN-based computer vision systems.

11.10 Conclusion

11.10.1 Summaries of the Cases Presented in This Chapter

In this chapter, we have looked at a few case studies, some of them using graph convolutional networks (GCNs), used in different computer vision applications. In looking at case studies, we focused on image classification, object identification and localization; segmentation with semantics; 3D vision; understanding video content, such as scenes from the University of Dubuque's composing class; or shots that required heavy redaction, like those in spy movies, in which they just had to use gibberish feeds. Every case study taught us something about the benevolence of GCNs in overcoming particular quagmires and improving computer vision performance.

11.10.2 The Importance of GCNs for Computer Vision Applications is Emphasized

The case studies featured in this section show how GCNs could steer computer vision applications forward. The use of graph convolutions provides a solid foundation for capturing complex relationships as well as contextual information among visual data. GCNs have shown considerable potential in a variety of image-processing tasks by taking advantage of graph structures. They greatly improve accuracy, robustness, and semantic understanding over the existing methods reported therein (see Section 11.5 in this paper for details). The case studies detailed above demonstrate that GCNs are a technique of great promise within computer vision.

11.10.3 Inspiration for Further Research and Exploration in this Area

With the results and concepts from these case studies, it is clear that there is still much to do in GCNs for computer vision. The conclusion asks these experts and scholars to carry on their work in developing this field. In addition, we would like to highlight several areas in need of further research. For instance: devising more energy-efficient GCN architectures, smoothing out the knotty scalability problem and discovering new applications for use with this paradigm from hitherto unexplored fields and domains, and increasing both interpretability as well as explainability of these models. Consequently, if we are willing to see where this frontier in computer vision can lead us—as long as there is still hope for it—maybe GCN will continue its meteoric rise into the realm of advanced technology.

Therefore, the case studies presented in this section illustrate that graph convolutional networks have proven to be effective and promising for use on computer vision problems. GCNs have proved they can pick up on complex correlations and raise the profile of visual data. As researchers and practitioners, let us believe in the magic of GCNs. Let us keep researching what this wonderful technology has to offer to adorn computer vision even more.

Bibliography

1. Kipf, T.N. and Welling, M., Semi-Supervised Classification with Graph Convolutional Networks, in: *Proceedings of the International Conference on Learning Representations (ICLR)*, 2017.

2. Hamilton, W., Ying, Z., Leskovec, J., Inductive representation learning on large graphs, in: *Advances in Neural Information Processing Systems*, pp. 1024–1034, 2017.
3. Veličković, P., Cucurull, G., Casanova, A., Romero, A., Lio, P., Bengio, Y., Graph attention networks, in: *International Conference on Learning Representations*, 2018.
4. Ying, R., You, J., Morris, C., Ren, X., Hamilton, W.L., Leskovec, J., Hierarchical graph representation learning with differentiable pooling, in: *Advances in Neural Information Processing Systems*, pp. 4805–4815, 2018.
5. Gilmer, J., Schoenholz, S.S., Riley, P.F., Vinyals, O., Dahl, G.E., Neural message passing for quantum chemistry, in: *Proceedings of the 34th International Conference on Machine Learning*, vol. 70, pp. 1263–1272, 2017.
6. Li, Y., *et al.*, Graph Convolutional Networks for Breast Cancer Diagnosis in Mammographic Images. *Med. Image Anal.*, 70, 102123, 2022.
7. Gupta, P., *et al.*, Person Re-Identification Using Graph Convolutional Networks. *Proceedings of the IEEE/CVF Conference on Computer Vision and Pattern Recognition (CVPR)*, 2022.
8. Wang, J., *et al.*, Graph Convolutional Networks for Video-Based Human Action Recognition. *Pattern Recognit.*, 126, 108321, 2023.
9. Zhang, H., *et al.*, Scene Understanding with Graph Convolutional Networks. *Proceedings of the European Conference on Computer Vision (ECCV)*, 2021.
10. Wang, H. and Schmid, C., Action recognition with improved trajectories, in: *Proceedings of the IEEE International Conference on Computer Vision*, pp. 3551–3558, 2013.
11. Simonyan, K. and Zisserman, A., Two-stream convolutional networks for action recognition in videos, in: *Advances in Neural Information Processing Systems*, pp. 568–576, 2014.
12. Tran, D., Bourdev, L., Fergus, R., Torresani, L., Paluri, M., Learning spatiotemporal features with 3D convolutional networks, in: *Proceedings of the IEEE International Conference on Computer Vision*, pp. 4489–4497, 2015.
13. Carreira, J. and Zisserman, A., Quo vadis, action recognition? A new model and the kinetics dataset, in: *Proceedings of the IEEE Conference on Computer Vision and Pattern Recognition*, pp. 6299–6308, 2017.
14. Yue-Hei Ng, J., Hausknecht, M., Vijayanarasimhan, S., Vinyals, O., Monga, R., Toderici, G., Beyond short snippets: Deep networks for video classification, in: *Proceedings of the IEEE Conference on Computer Vision and Pattern Recognition*, pp. 4694–4702, 2015.
15. Donahue, J., Hendricks, L.A., Guadarrama, S., Rohrbach, M., Venugopalan, S., Saenko, K., Darrell, T., Long-term recurrent convolutional networks for visual recognition and description, in: *Proceedings of the IEEE Conference on Computer Vision and Pattern Recognition*, pp. 2625–2634, 2015.
16. Li, Y., Song, Y., Li, J., Huang, T., Temporal modeling of dynamic scenes for action recognition. *IEEE Trans. Image Process.*, 26, 3, 1145–1158, 2017.

17. Ren, S., *et al.*, Faster R-CNN: Towards real-time object detection with region proposal networks, in: *Advances in Neural Information Processing Systems*, pp. 91–99, 2015.
18. He, K., *et al.*, Mask R-CNN, in: *Proceedings of the IEEE International Conference on Computer Vision (ICCV)*, pp. 2961–2969, 2017a.
19. William, P., Yogeesh, N., S. Vimala, Gite, P., Selva, K.S., Blockchain Technology for Data Privacy using Contract Mechanism for 5G Networks. *2022 3rd International Conference on Intelligent Engineering and Management (ICIEM)*, London, United Kingdom, pp. 461–465, 2022, doi: 10.1109/ICIEM54221.2022.9853118.
20. Mishra, S., Choubey, S., Choubey, A., Yogeesh, N., Durga Prasad Rao, J., William, P., Data Extraction Approach using Natural Language Processing for Sentiment Analysis. *2022 International Conference on Automation, Computing and Renewable Systems (ICACRS)*, Pudukkottai, India, pp. 970–972, 2022, doi: 10.1109/ICACRS55517.2022.10029216.
21. William, P., Yogeesh, N., Tidake, V.M., Sumit Gondkar, S., Chetana, R., Vengatesan, K., Framework for Implementation of Personality Inventory Model on Natural Language Processing with Personality Traits Analysis. *2023 International Conference on Intelligent Data Communication Technologies and Internet of Things (IDCIoT)*, Bengaluru, India, pp. 625–628, 2023, doi: 10.1109/IDCIoT56793.2023.10053501.
22. Gupta, K., Choubey, S., Yogeesh, N., William, P., Vasanthakumari, T.N., Kale, C.P., Implementation of Motorist Weariness Detection System using a Conventional Object Recognition Technique. *2023 International Conference on Intelligent Data Communication Technologies and Internet of Things (IDCIoT)*, Bengaluru, India, pp. 640–646, 2023, doi: 10.1109/IDCIoT56793.2023.10052783.

12

Case Study and Use Cases of Dynamic Graphs in GCNN for Computer Vision

S. Anubha Pearline* and **S. Geetha**

School of Computer Science and Engineering, Vellore Institute of Technology, Chennai Campus, Chennai, India

Abstract
In recent years, deep learning has been considered lucrative in various domains such as computer vision and natural language processing. Apart from deep learning, Graph Neural Networks (GNNs) have captured a lot of interest among researchers. Exclusively for computer vision applications, graph convolutional neural network (GCNN) is widely used. GNN and GCNN rely on graph data-based neural networks. Convolutional neural networks (CNN) can have images as direct inputs, while GCNN reveals the image grids or patches as graph nodes. Many different GNNs and three various applications of computer vision are discussed in this book chapter.

Keywords: Graph convolutional neural networks, computer vision, dynamic graphs, convolutional neural networks

12.1 Introduction

Over the past decade, deep learning has risen and emerged as an outstanding technique in Artificial Intelligence (AI) and Machine learning (ML). The major highlight of deep learning is that it extracts a lot of innate features from images in Computer Vision (CV) applications [1]. Image processing or CV encompasses dataset acquisition, image pre-processing, extraction of features, and image classification [2]. Graph Convolutional Neural Networks (GCNNs) depend on graphs, nodes, and edges.

**Corresponding author*: anubhapearline.s@vit.ac.in; anubhapearl@gmail.com

Malini Alagarsamy, Rajesh Kumar Dhanaraj, J. Felicia Lilian, Vandana Sharma and Gheorghita Ghinea (eds.) Graph Convolutional Neural Networks for Computer Vision, (255–274) © 2026 Scrivener Publishing LLC

According to GCNNs, images are split into several image patches or grids. Explainability and detailed visualization of images are less explored using GCNNs. Initially, GNNs using graph concept was introduced mainly for social network analysis, finance, etc. [3]. Since the inception of GCNNs, it has been analyzed over computer vision applications such as handwritten image recognition [4], object detection [5], and video segmentation [6].

Autonomous vehicles, medical image analysis, facial recognition, augmented reality, industrial automation, satellite image analysis, and plenty of other areas use computer vision. Large datasets, improved technology, and developments in deep learning approaches have all contributed to the recent substantial improvements in computer vision. As a result, it will continue to be extremely important in determining the direction of technology and AI. Handcrafted feature-classification algorithms are used to address the problems in CV. Convolutional neural networks are currently experiencing rising attention for use in recognizing images. Graph Neural Networks (GNN) or Graph Convolution Neural Networks (GCNNs) [1–6] are a boon to CV researchers. GCNNs are less concentrated in the arena of CV applications. Figure 12.1 depicts the CV techniques present.

A system has to learn from graph-based inputs to resemble physical systems, understand molecular fingerprints, predict interactions among proteins, and classify diseases. Graph reasoning models are also needed in various fields, including acquiring knowledge from non-structural data, like texts and images, where reasoning on extracted structures based on gathered images is a prominent topic of research. GCNNs are neural network models that rely on the transmission of messages between nodes in a graph to capture and simulate the dependencies and relationships inherent in the graph structure [7].

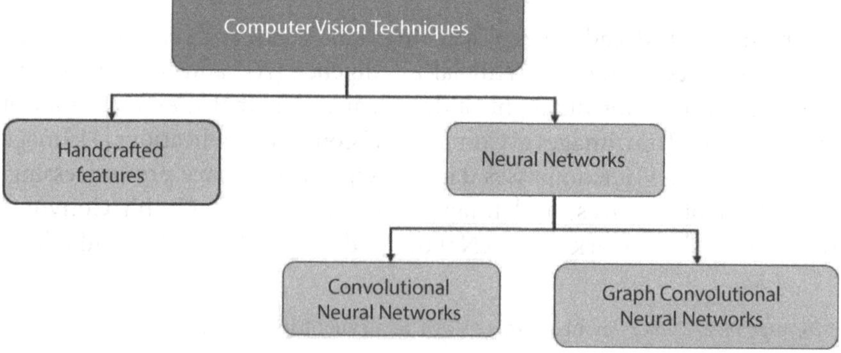

Figure 12.1 Different techniques used by computer vision applications.

This book chapter focuses on various image classification applications related to graph neural networks. In Subsections 12.1.1 and 12.1.2, CV approaches, such as handcrafted features and convolutional neural networks, are addressed. Section 12.2 contains the graph convolutional neural network. Section 12.3 discusses the case studies in graph neural networks. Section 12.4 details the challenges and issues faced by GCNN-based CV applications. Section 12.5 concludes this book chapter.

12.1.1 Handcrafted Features

The main elements involved in the conventional method for image processing or computer vision encompass dataset acquisition, image preprocessing, feature extraction, and image classification. Figure 12.2 demonstrates the procedure of image classification based on handcrafted features. From the dataset, each input image is preprocessed. Further, features are extracted and classified using algorithms.

Dataset Collection: Gathering images through cameras, sensors, or other imaging devices is the first step in CV. These devices capture images or video frames. The captured images are further processed using computer algorithms.
Image Preprocessing: Preprocessing techniques are used to enhance the quality of images, remove noise, and prepare the data for analysis. Necessarily, the images are resized using CV methods present in the OpenCV tool (if Python is used).
Image Segmentation: This is an optional step involved in image classification. Image segmentation is the segmentation only of a part of an object or the segmentation of important information present in the image.
Feature Extraction: CV algorithms identify and extract relevant features from images, such as edges, corners, textures, or object key points. Many feature extraction algorithms, such as Hu moments, Haralick texture, and local binary pattern (LBP), are present. These features serve as the basis for further analysis. Extracted features will be a set of values ordered as a 1D vector.

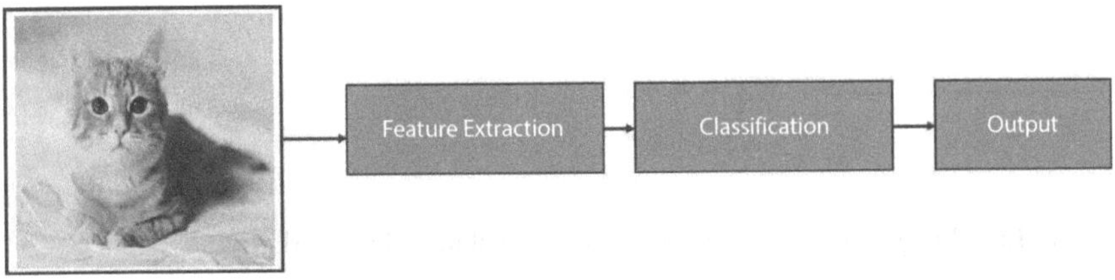

Figure 12.2 Image classification using handcrafted features.

Classification: It is a major step in CV to classify images based on features obtained in the feature extraction step. Classification is performed using machine learning methods [8], including support vector machines (SVM), classification and regression trees (CART), the Naïve Bayes algorithm, and random forest.

12.1.2 CNN

A Convolutional Neural Network (CNN) [9], a specialized type of neural network, is capable of analyzing images or videos. However, CNNs also have the capability of working well for non-image data. CNNs consist of convolutional layers, pooling layers, and fully connected layers. CNN is used for the recognition of plant species. Figure 12.3 shows the processing of CNN using images.

Convolution Image

Convolution is mathematically defined as the sum of the element-wise products of two matrices. This convolution operation uses the Toeplitz matrix [9]. Consider the image "X" and filter "Y." X and Y are both matrices, with matrix X representing image pixels. Upon convolving the image "X" with the filter "Y," we derive a matrix of feature maps denoted as "Z."

A filter assesses the similarity between a segment of the input and a specific feature, such as a diagonal, an arch, or an edge (whether vertical or horizontal). Essentially, a filter acts as a unique template or pattern that, when convolved across an input image, searches for resemblances between various areas of the input and the predefined template.

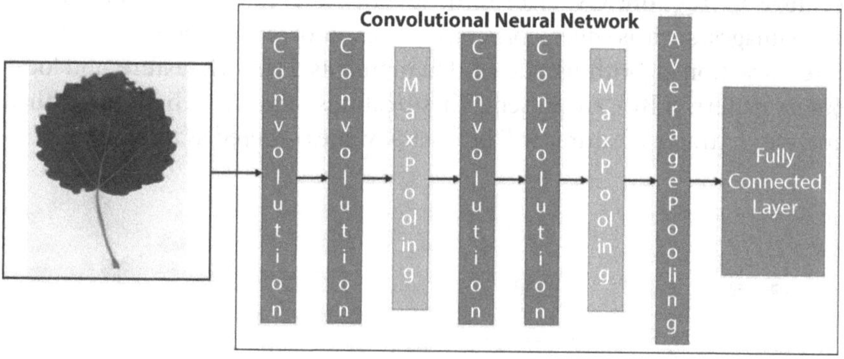

Figure 12.3 Image classification using convolutional neural network.

Padding: The padding parameter is used to maintain the image's dimension consistently. In convolution, the stride parameter dictates the rate at which the filter traverses the pixels of the input image along both the vertical and horizontal axes. Stride is ultimately determined by how we expect to see in the final image. We choose a smaller stride size if we anticipate that our output will exhibit several fine-grained characteristics. We shall select a larger stride size if the macro-level of features is our only concern.
Pooling Layer: Another fundamental building element in CNN is the pooling layer. It attempts to determine whether or not a certain region in the image contains the characteristic in question. The pooling layer examines more significant portions of the image (those with several patches) and collects aggregate statistics (min, max, average, and global). This leaves the network resistant to adjustments that occur locally.
Fully Connected Layers help perform classification. At least one fully connected layer [9] is typically included employing a SoftMax activation function.

12.2 Graph Convolutional Neural Networks (GCNNs)

Graph: Graph Convolutional Neural Networks (GCNNs) [1–6], as the name suggests, are constructed based on graphs. A graph structure contains several nodes or vertices connected through edges. A graph comprises two fundamental elements: vertices (also known as nodes) and edges. It is represented as G = (V, E), where V denotes the set of vertices, and E denotes the set of edges connecting these vertices.

Edges are considered directed when there is a directional dependency among the nodes. If edges connected to nodes do not have head pointers, then edges are undirected. Graph, G, considers the usage of adjacency matrix and information from neighboring nodes. Hence, the related concepts are considered under Subsections 12.2.1.1 and 12.2.1.2.

Graphs are rapidly gaining prominence in machine learning due to their extremely powerful representational properties. Fine-grained relationships are essential for perceiving, comprehending, and reasoning with visual input. Compared to convolutional neural networks, GCNNs can extract more detailed information. To simplify the relevant CV tasks, GCNNs can be naturally used to extract patterns from these graphs [10]. An embedding is associated with every GCNN node. The exact position of the node in the data space is conveyed by this embedding. Neural network structures that communicate with graphs are known as GCNNs. The primary goal of a GCNN architecture is to generate an embedding that includes data about its neighborhood. Multiple issues can be resolved with an embedding, such as edge prediction and node labeling.

In summary, GCNNs represent a subclass of deep learning algorithms engineered to execute interpretation tasks on data structured in graph format. They are used in graph-based models and can make predictions at the node, edge, and overall graph levels. Many CV applications, such as anomaly detection and object detection, focus on graph neural networks for solutions.

12.2.1 Graph and Images

An image on its own is a graph. Figure 12.4 exhibits how the image might be expressed as a grid graph. For both internal and corner nodes, the count of outgoing edges from each node remains consistent. Due to this uniformity, the grid graph can easily undergo methods similar to convolution. Every pixel in an image operates as a node connected to neighboring pixels by hypothetical edges forming a specific graph.

Let us imagine images as rectangular grids containing color image channels as in Figure 12.5. These images can be interpreted as pixel arrays (e.g., 5 × 5 × 3 floats). Furthermore, images can be conceptualized as graphs possessing a uniform structure, wherein each pixel within the image corresponds to a node. Every node is connected to adjacent pixels through an edge. Every non-border pixel has eight neighbors, and each node retains a three-dimensional vector that contains the RGB value of each pixel.

12.2.1.1 Graph Nodes

Graph nodes are inferred for computer vision tasks depending on the nature of the input, whether it be an image or a video. Each node in the graph has associated features (vectors). These features embody the features or attributes of the nodes. For instance, in a social network, nodes might represent individuals, with features encompassing demographic information.

Graph nodes make use of adjacency matrix and neighborhood node information for processing. An adjacency matrix is generated based on the presence of an edge connecting two nodes denoted as and j. It is represented as follows:

$$A_{ij} = \begin{cases} 1, \textit{if E exists from } i \rightarrow j \\ 0, \textit{else} \end{cases} \tag{12.1}$$

Figure 12.4 Image patches for graph nodes.

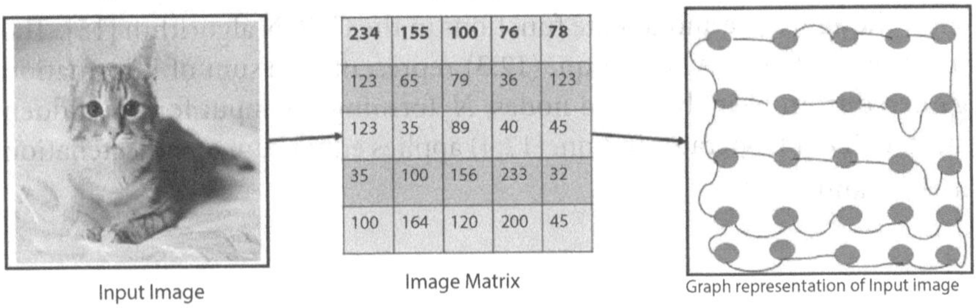

Figure 12.5 Graph representation for the given input image.

12.2.1.2 Graph Edges

Edges in the graph can also have associated features. These features capture the relationships or interactions between nodes. In a social network, edge features could represent friendship strength or communication frequency. Graph edges can be classified as spatial, spectral, and temporal edges. The spatial domain treats convolution operations in spatial regions similar to ChebNet [11].

12.2.1.3 Message Passing

The core operation in GNNs is message passing or information propagation. Hidden embedding or hidden layer utilizes this feature. GNNs aggregate data from nearby nodes and edges to iteratively update node features [12]. This process allows nodes to learn from their neighbors. This message-passing concept is adopted by various networks including graph convolutional networks (GCNs) [13], GraphSAGE [13], and graph attention networks (GATs) [14]. If $u \in V$, then the simple layer of GNN is represented by the following:

$$f_k = \sigma\left(\sum_i \sum_j V_{ij} w_{jk}\right) \tag{12.2}$$

In Equation 12.2, σ represents the activation function. V_{ij} denotes the node feature, and w_{jk} represents the trainable weights.

12.2.1.4 Aggregation Function

An aggregation function aggregates data from neighboring nodes and edges. This function is depicted in Figure 12.6. Aggregation functions that are frequently employed include mean, maximum, and weighted sum. In CNN, we typically employ average or maximum pooling. GNN or GCNN, like CNN, uses an aggregation mechanism. Equations (12.3) and (12.4) show the aggregate and update functions in the GCNN algorithm [15]. The function AGGREGATE in Eqn. (12.3) represents the sum of information gathered from Neighborhood nodes, N_v forming the input for the hidden layer. UPDATE operator in Eqn. (12.4) applies either sum or concatenation to $h_v^{(l+1)}$ and $m_v^{(l+1)}$.

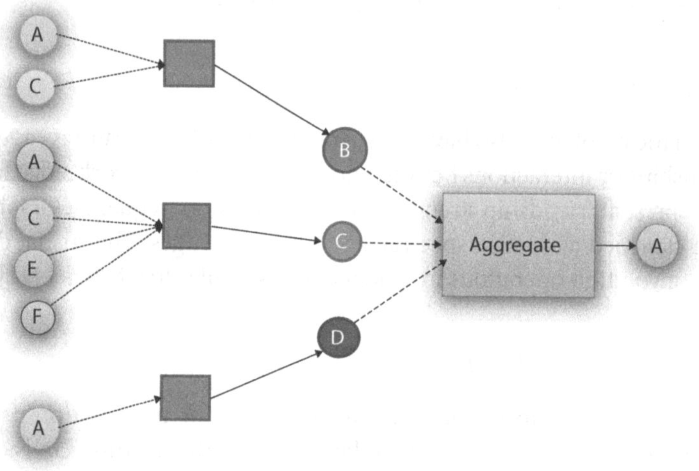

Figure 12.6 Aggregation function in GCNN.

$$m_v^{(l+1)} = AGGREGATE\left(\left\{h_n^{(l)} \mid u \in N_v\right\}\right) \tag{12.3}$$

$$h_v^{(l+1)} = UPDATE\left(h_v^{(l)}, m_v^{(l+1)}\right) \tag{12.4}$$

12.2.1.5 Graph Embeddings

Graph embeddings reduce the feature representations from high dimensions to low dimensions.

12.2.1.6 Neural Network Layers

GCNNs typically consist of multiple layers, and each layer performs message passing followed by a neural network operation such as a fully connected layer or a convolutional layer. The learned representations become increasingly refined with each layer.

12.2.1.7 Output

Depending on the objective, the GNN's last layer might generate node- or graph-level outputs. As an illustration, in a node classification task, the GNN may generate a label for each node.

12.2.2 Types of Graph Convolutional Neural Networks

The main benefit of GCNN is its capacity to perform tasks that CNNs (convolutional neural networks) cannot manage. Multiple neighborhood node information is collectively gathered for further processing. Very complicated tasks can be handled with the help of GCNN. Hence, convolutional neural networks are employed for tasks involving recognition, object identification, and image classification. CNN incorporates pooling layers and hidden convolutional layers for classification.

A fundamental principle of GCNN is that nodes are assigned by their connections and neighbors. To understand this, imagine that a node would lose all of its information if all of its neighbors were eliminated. Consequently, a node is described by its neighbors and the relationships between them.

Graphs offer a straightforward means of describing data with intricate associations like those seen in recommendation systems, biological structures, and social networks. The primary goal of GCNNs is to learn and extract useful information from graph-structured data. GNNs extend the principles of traditional neural networks, which operate on structured data like grids (e.g., images) or sequences (e.g., text). Instead of fixed grid-like structures, GNNs operate on irregular and interconnected data.

The various GCNN architectures available are graph convolutional networks (GCNs) [16], GraphSAGE [17], gated graph convolutional neural networks (GGNNs) [19], and graph attention networks (GAT) [14], etc. GCNNs [1–6] have significantly impacted fields including social network analysis, biology, and recommendation systems. They persist as an emerging domain of inquiry within the realms of machine learning and artificial intelligence.

GCNNs act as an influential and significant method for performing predictive tasks at the node, edge, and graph levels [1–6]. GCNNs are types of neural networks devised specifically for more interpretable graph structures. There are several varieties of graph convolutional neural networks, which will be discussed in more detail in the sections that follow.

12.2.2.1 Spectral Graph Convolutional Network

The spectral graph convolutional network utilizes eigenvalues *via* the Laplacian matrix [16]. It utilizes convolution operation over graph-structured data. ChebNet is another architecture that utilizes Chebysev polynomial approximation.

12.2.2.2 GraphSAGE

Dynamic spatial and temporal characteristics can be simultaneously captured using GraphSAGE [17]. The architecture comprises dilated casual convolution, attention mechanism, and graph embeddings. The attention mechanism present in GraphSAGE can learn the weights dynamically.

12.2.2.3 Multiscale Dynamic Graph Convolutional Network (MDGCN)

MDGCN handles dynamic graph-structured data. This methodology combines multiscale dynamic convolution of input images and GCNN. For multiscale dynamic convolution, multiple input graphs are considered.

Various graphs with diverse neighborhood scales are employed to incorporate hidden spatial context across distinct scales.

12.2.2.4 Spectral–Spatial Graph Convolutional Networks (S^2-GCNs)

S^2-GCN [18] is an image-specific semi-supervised learning framework. Similar to graph convolutional network, spectral GCN incorporates the Laplacian operator. Furthermore, it is extremely dependent on the adjacent matrix of nodes.

12.2.2.5 Graph Attention Networks (GATs)

GAT [14] is a novel convolutional neural network based on graph structure. It contains a self-attention layer for handling the features of images. The attention mechanism utilizes an inductive learning approach.

12.2.2.6 Gated Graph Convolutional Neural Networks

Gated graph convolutional neural networks [19] is an integrated concept involving both gated recurrent unit (GRU) and graph convolution networks (GCNs).

12.2.2.7 Graph AutoEncoders (GAEs)

Unsupervised learning techniques associated with GAEs [20] encode graphs or nodes in a latent vector space. Afterward, the encoded data are used to recreate the graph. Graph autoencoders (GAEs) are employed to identify patterns embedded within networks and those involved in graph generation. GAEs create latent representations of nodes for network embedding by reconstructing structural graph elements, such as the adjacency matrix. Certain graph construction techniques develop nodes and edges incrementally, while others construct the entire graph at once.

12.3 GCNN Case Studies

Applications using GCNNs are discussed in this section including image classification, object detection, segmentation of images, and scene understanding.

12.3.1 Image Classification

Based on the trained dataset, image classification assigns a label or category to an entire test image. Convolutional neural networks (CNNs), a prominent deep learning methodology, have significantly enhanced the precision of image classification. In this subsection, various applications in image classification utilizing graph convolution networks are discussed.

12.3.1.1 Hyperspectral Image Classification (HSI)

Kavran *et al.* [21] proposed the utilization of GraphSAGE and long short-term memory (LSTM) for node classification. Features are extracted from spatiotemporal images using EfficientNetV2-CNN. Yao *et al.* [22] introduced a deep hybrid multi-graph neural network (DHMG) for HSI, illustrated in Figure 12.7. The DHMG has two distinct units. In the first branch, the spectral graph filter extracted spectral features. In the second branch, an autoregressive moving average (ARMA) filter was applied to reduce noise within the graph. To extract and fine-tune graph features, the GraphSAGE network is shown.

Wan *et al.* [23] developed the multiscale dynamic graph convolutional network (MDGCN) to discriminate hyperspectral images. MDGCN was designed to overcome the restrictions, such as square pixels, which are considered for processing in CNN. Multiple input graphs at various scales are identified and processed.

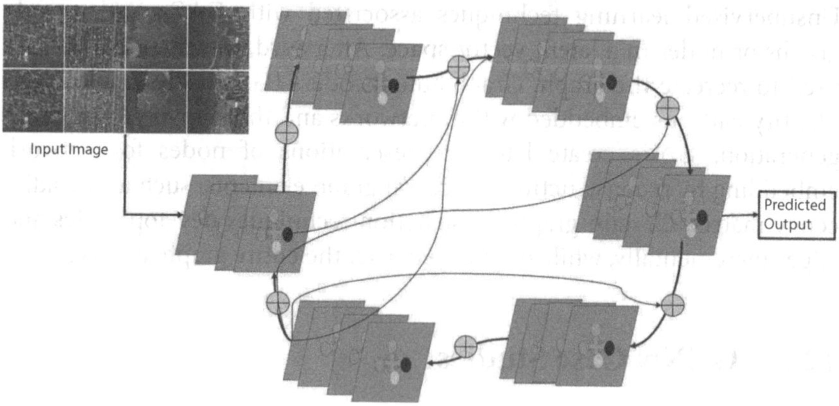

Figure 12.7 Architecture for DHMG.

Qin *et al.* [18] developed spectral–spatial graph convolutional networks (S^2-GCNs) for the classification of hyperspectral images. Complete spatial information is considered for the current pixel. Sha *et al.* [24] introduced graph attention networks (GATs), a type of graph neural network variant, for classifying hyperspectral images that include both labeled and unlabeled datasets. Based on attention coefficients, weights are assigned to the graph convolution network. Over the labeled and unlabeled samples, a graph model is constructed using spatial–spectral measures.

12.3.1.2 Breast Cancer Detection

Computer-aided diagnosis is very crucial for breast histopathological images. Gao *et al.* [25] proposed an integrated version of CNN and GCN for classifying breast cancer images. The methodology utilized for analyzing breast cancer histopathological images is depicted in Figure 12.8.

A multi-modal GCN architecture for breast cancer prediction is created by Palmal *et al.* [26]. Using GCN, features are retrieved and concatenated together. After that, the collected features are fed into a stack ensemble model. Zhang *et al.* [27] demonstrated a CNN that incorporates batch normalization (B), dropout (D), and rank-based stochastic pooling (R). The name of this CNN is BDR-CNN. Using a two-layer GCN, this model has been hybridized.

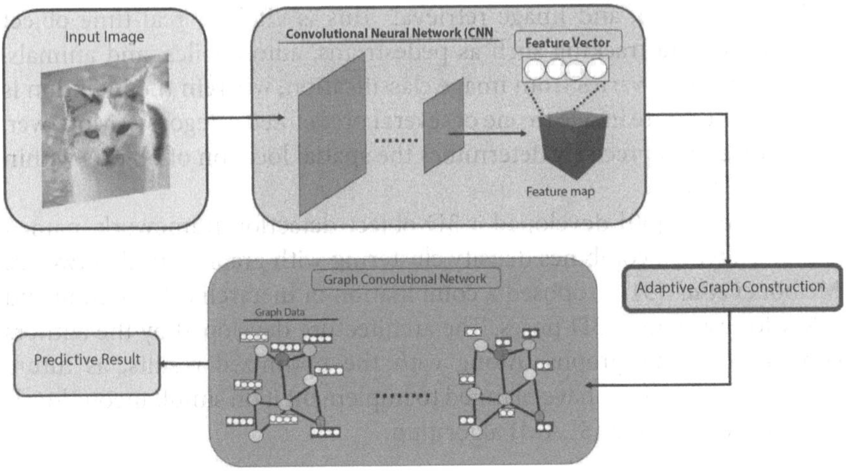

Figure 12.8 Combined model of CNN and GCN for breast cancer classification (Gao *et al.*).

12.3.1.3 Coronary Heart Disease Prediction

Xue *et al.* [28] introduced adaptive multi-channel graph convolutional neural networks (AM-GCNs), which aggregate information from both graph embeddings and node representations. Together with the AM-GCN, the authors additionally included a population graph, a K-nearest neighbor approach, and an attention mechanism.

Lin *et al.* [29] presented a domain-adaptive multichannel graph convolutional network (DAMGCN) for coronary heart disease. The authors used a two-channel GCN and combined their output features. For source embedding, node representations are identified. Additionally, an attention mechanism is introduced to create a graph representation. Optimization minimizes the differences between the classifiers in the source and target domains. Three loss functions are employed for this analysis. The first loss function present in DAMGCN is the loss source classifier. Second, the loss function is used for adversarial domains inspired by GAN. Last, the loss function for the target classifier is used. Transfer learning of GCN is applied for predicting coronary heart disease.

12.3.2 Object Detection

Object detection entails the identification and categorization of objects within an image or video frame. It is a fundamental problem in CV that has multiple practical applications including self-driving cars, surveillance, facial recognition, and image retrieval. This is vital for real-time object recognition and tracking such as pedestrians, automobiles, and animals. Object detection varies from image classification, wherein the intention is to assign the entire image to one of several predefined categories. Moreover, object detection precisely determines the spatial location of objects within an image [10].

Xiong *et al.* [30] developed a 3D object detection framework, named DCGNN, which combines density clustering with graph neural networks. Melbouci *et al.* [31] proposed a combination of hierarchical attention and GNN for estimating 3D poses. The architecture developed by the authors contains dynamic graphs. Along with the optimized results, as future direction, the authors have planned to implement it on simultaneous localization and mapping (SLAM) algorithm.

12.3.3 Image Segmentation and Scene Understanding

The technique of splitting an image into distinct elements or portions according to shared properties, such as texture, color, or intensity, is known as image segmentation. The benefit of graph convolutional networks is that they can be used and tested on applications such as medical image analysis and scene understanding containing moving objects.

GCNNs are proficient neural networks that can represent both local and global regions resulting in comprehensive context-aware semantic understanding. Scene understanding is determining the geometric and semantic cues present in an image or video for developing an extensive awareness of visual content [32]. Scene understanding can be defined otherwise as, "To examine a scene by taking into consideration the inherent relationships between its contents and their geometric and semantic context" [32].

Wang *et al.* [33] devised the attentive GCNN to address the challenge of zero-shot video object segmentation (ZVOS). Qi *et al.* [34] introduced a three-dimensional graph neural network (3D GNN), which utilizes 3D point cloud data and k-nearest neighbor graph techniques. The hidden node representations of GNNs are dynamically updated demonstrating effectiveness comparable to recurrent neural networks (RNNs). In addition, they used a pre-trained very large convolutional neural network (VGG-16) in their strategy.

Zhang *et al.* [35] designed a relation-based GCN for panorama 3D scene understanding, and it is shown in Figure 12.9. Liu *et al.* [36] introduced

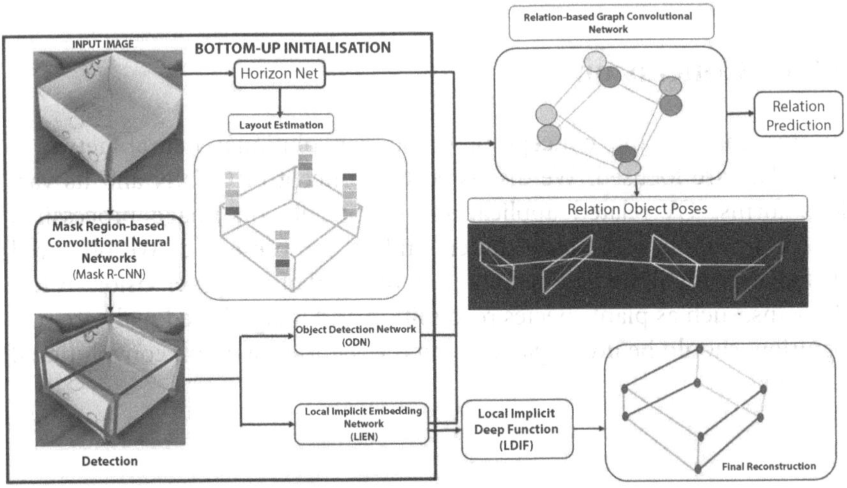

Figure 12.9 Process diagram of relation-based GCN [34].

the self-constructing graph network (SCG-Net) for semantic segmentation utilizing a blend of Kullback–Leibler divergence regularization and graph-reconstruction terms. SCG-Net assumed the semantic segmentation task as a node classification problem.

12.4 Challenges and Issues in GCNN for CV

One of the primary issues in image classification is the scarcity of labeled images. Image datasets are significantly smaller in size. As a result, the majority of the results require data augmentation. Data augmentation is still constrained by hardware and software constraints. A tedious procedure is the development of graphs followed by classification. For graph convolutional networks, the spatial information contained in images is of greater importance. Tracing time-based visuals (for example, moving objects in films) as well as generating precise predictions using the graph notion is a significant challenge.

Time complexity is a reminiscent factor considering GCNN. The temporal complexity of each variant of GCNN is swayed with respect to the number of nodes (N) and number of edges (E) formed from graph representation. For instance, graph traversal has a time complexity of $O(|N|+|E|)$. GCNN employed sets of matrix multiplications. Hence, the computational complexity of GCNN is $O(N^3)$. Depending on the type of GCNN employed, minimizing the number of nodes may decrease the cost of the algorithm.

12.5 Conclusion

In this chapter, the concept of graph convolutional neural networks (GCNNs) are focused. We discussed the concept of GCNN and its various forms. CV-related applications specifically on image processing, object detection, and scene understanding are presented. We observed that GCNN is still being unexplored in the field of computer vision. A few domains, such as plant species recognition and image-based malware recognition, should be investigated using the theories of graph convolutional networks.

References

1. Zhang, Z., Cui, P. and Zhu, W., Deep learning on graphs: A survey. *IEEE Trans. Knowl. Data Eng.*, 34, 1, 249–270, 2020.
2. Nanni, L., Ghidoni, S., Brahnam, S., Handcrafted vs. non-handcrafted features for computer vision classification. *Pattern Recognit.*, 71, 158–172, 2017.
3. Yuan, H., Yu, H., Gui, S., Ji, S., Explainability in graph neural networks: A taxonomic survey. *IEEE Trans. Pattern Anal. Mach. Intell.*, 45, 5, 5782–5799, 2022.
4. Kajla, N.I., Missen, M.M.S., Luqman, M.M., Coustaty, M., Graph neural networks using local descriptions in attributed graphs: an application to symbol recognition and hand written character recognition. *IEEE Access*, 9, 99103–99111, 2021.
5. Li, X., Zhao, L., Wei, L., Yang, M.H., Wu, F., Zhuang, Y., Ling, H., Wang, J., Deepsaliency: Multi-task deep neural network model for salient object detection. *IEEE Trans. Image Process.*, 25, 8, 3919–3930, 2016.
6. Brissman, E., Johnander, J., Danelljan, M., Felsberg, M., Recurrent Graph Neural Networks for Video Instance Segmentation. *Int. J. Comput. Vision*, 131, 2, 471–495, 2023.
7. Zhou, J., Cui, G., Hu, S., Zhang, Z., Yang, C., Liu, Z., Wang, L., Li, C., Sun, M., Graph neural networks: A review of methods and applications. *AI Open*, 1, 57–81, 2020. https://doi.org/10.1016/j.aiopen.2021.01.0012.
8. Sundara Sobitha Raj, A.P. and Vajravelu, S.K., DDLA: dual deep learning architecture for classification of plant species. *IET Image Proc.*, 13, 12, 2176–2182, 2019.
9. Goodfellow, I., Bengio, Y., Courville, A., *Deep learning*, vol. 1, no. 2, Cambridge MIT Press, 2016.
10. Tang, S., Zhang, W., Mu, Z., Shen, K., Li, J., Li, J., Wu, L., Graph Neural Networks in Computer Vision, in: *Graph Neural Networks: Foundations, Frontiers, and Applications*, pp. 447–462, 2022.
11. Tang, S., Li, B., Yu, H., ChebNet: efficient and stable constructions of deep neural networks with rectified power units *via* Chebyshev approximation. *Commun. Math. Stat.*, 1–27, 2024.
12. Ward, I. R., Joyner, J., Lickfold, C., Guo, Y., Bennamoun, M., A Practical Tutorial on Graph Neural Networks. *ACM Comput. Surv.*, 54, 10, 2022, https://doi.org/10.1145/3503043.
13. Xu, K., Hu, W., Leskovec, J., Jegelka, S., *How powerful are graph neural networks?*, 2018, *arXiv preprint arXiv:1810.00826*.
14. Veličković, P., Cucurull, G., Casanova, A., Romero, A., Lio, P., Bengio, Y., *Graph attention networks*, 2017, *arXiv preprint arXiv:1710.10903*.
15. Lachaud, G., Conde-Cespedes, P., Trocan, M., Mathematical Expressiveness of Graph Neural Networks. *Mathematics*, 10, 24, 2022.

16. Baroud, S., Chokri, S., Belhaous, S., Mestari, M., A brief review of graph convolutional neural network based learning for classifying remote sensing images. *Procedia Comput. Sci.*, 191, 349–354, 2021.
17. Liu, J., Ong, G.P., Chen, X., GraphSAGE-based traffic speed forecasting for segment network with sparse data. *IEEE Trans. Intell. Transp. Syst.*, 23, 3, 1755–1766, 2020.
18. Qin, A., Shang, Z., Tian, J., Wang, Y., Zhang, T., Tang, Y.Y., Spectral–spatial graph convolutional networks for semisupervised hyperspectral image classification. *IEEE Geosci. Remote Sens. Lett.*, 16, 2, 241–245, 2018.
19. Beck, D., Haffari, G., Cohn, T., *Graph-to-sequence learning using gated graph neural networks*, 2018, 2018, arXiv preprint arXiv:1806.09835.
20. Wu, Z., Pan, S., Chen, F., Long, G., Zhang, C., Philip, S.Y., A comprehensive survey on graph neural networks. *IEEE Trans. Neural Netw. Learn. Syst.*, 32, 1, 4–24, 2020.
21. Kavran, D., Mongus, D., Žalik, B., Lukač, N., Graph Neural Network-Based Method of Spatiotemporal Land Cover Mapping Using Satellite Imagery. *Sensors*, 23, 14, 6648, 2023.
22. Yao, D., Zhi-li, Z., Xiao-feng, Z., Wei, C., Fang, H., Yao-ming, C., Cai, W.W., Deep hybrid: multi-graph neural network collaboration for hyperspectral image classification. *Def. Technol.*, 23, 164–176, 2023.
23. Wan, S., Gong, C., Zhong, P., Du, B., Zhang, L., Yang, J., Multiscale dynamic graph convolutional network for hyperspectral image classification. *IEEE Trans. Geosci. Remote Sens.*, 58, 5, 3162–3177, 2019.
24. Sha, A., Wang, B., Wu, X., Zhang, L., Semisupervised classification for hyperspectral images using graph attention networks. *IEEE Geosci. Remote Sens. Lett.*, 18, 1, 157–161, 2020.
25. Gao, Z., Lu, Z., Wang, J., Ying, S., Shi, J., A convolutional neural network and graph convolutional network based framework for classification of breast histopathological images. *IEEE J. Biomed. Health. Inf.*, 26, 7, 3163–3173, 2022.
26. Palmal, S., Arya, N., Saha, S., Tripathy, S., A Multi-modal Graph Convolutional Network for Predicting Human Breast Cancer Prognosis, in: *International Conference on Neural Information Processing*, Springer Nature Singapore, Singapore, pp. 187–198, 2022.
27. Zhang, Y.D., Satapathy, S.C., Guttery, D.S., Górriz, J.M., Wang, S.H., Improved breast cancer classification through combining graph convolutional network and convolutional neural network. *Inf. Process. Manage.*, 58, 2, 102439, 2021.
28. Xue, Y., Chen, K., Lin, H., Zhong, S., All-Cause Death Prediction Method for CHD Based on Graph Convolutional Networks. *Comput. Intell. Neurosci.*, 2389560, 1–14, 2022.
29. Lin, H., Chen, K., Xue, Y., Zhong, S., Chen, L., Ye, M., Coronary heart disease prediction method fusing domain-adaptive transfer learning with graph convolutional networks (GCN). *Sci. Rep.*, 13, 1, 14276, 2023.

30. Xiong, S., Li, B., Zhu, S., DCGNN: A single-stage 3D object detection network based on density clustering and graph neural network. *Complex Intell. Syst.*, 9, 3, 3399–3408, 2023.
31. Melbouci, K. and Nashashibi, F., Hierarchical Attention and Graph Neural Networks: Toward Drift-Free Pose Estimation, 2023, http://arxiv.org/abs/2309.09934.
32. Naseer, M., Khan, S., Porikli, F., Indoor scene understanding in 2.5/3d for autonomous agents: A survey. *IEEE Access*, 7, 1859–1887, 2018.
33. Wang, W., Lu, X., Shen, J., Crandall, D.J., Shao, L., Zero-shot video object segmentation *via* attentive graph neural networks, in: *Proceedings of the IEEE/CVF International Conference on Computer Vision*, pp. 9236–9245, 2019.
34. Qi, X., Liao, R., Jia, J., Fidler, S., Urtasun, R., 3D graph neural networks for RGBD semantic segmentation, in: *Proceedings of the IEEE International Conference on Computer Vision*, pp. 5199–5208, 2017.
35. Zhang, C., Cui, Z., Chen, C., Liu, S., Zeng, B., Bao, H., Zhang, Y., Deeppanocontext: Panoramic 3d scene understanding with holistic scene context graph and relation-based optimization, in: *Proceedings of the IEEE/CVF International Conference on Computer Vision*, pp. 12632–12641, 2021.
36. Liu, Q., Kampffmeyer, M., Jenssen, R., Salberg, A.B., Self-Constructing Graph Convolutional Networks for Semantic Labeling, in: *Proceedings of the IGARSS 2020 - 2020 IEEE International Geoscience and Remote Sensing Symposium*, Waikoloa, HI, USA, 2021.

About the Editors

Dr. Malini Alagarsamy is currently serving as an Associate Professor in School of Computer Science and Engineering at Vellore Institute of Technology, Chennai, Tamilnadu, India. She has 20+ years of teaching experience. She obtained her doctoral degree from Anna University, Chennai. She is a lifetime member of Computer Society of India. She has published 60+ research articles in refereed journals and international/national conferences. She has contributed 20+ chapters to book. She is holding two Indian Design patents. She has contributed 3 edited books on Artificial Intelligence and Machine Learning. Dr. Malini has contributed voluntarily as a reviewer for reputed international journals and session chair for international conferences. She acted as mentor and won smart India Hackathon 2022 held at NIT Assam. Her research interest includes software Engineering, Software Testing, Mobile Application development, Green Computing, Internet of Things, Artificial Intelligence and Machine Learning.

Dr. Rajesh Kumar Dhanaraj is a distinguished Professor at Symbiosis International (Deemed University) in Pune, India. Prior to this, he served as a Professor at the School of Computing Science & Engineering at Galgotias University in Greater Noida, India. His exceptional academic and research contributions have placed him among the top 2% of scientists globally, an honor recognized by Elsevier and Stanford University. Dr. Dhanaraj completed his B.E. in Computer Science and Engineering from Anna University Chennai, followed by an M.Tech from Anna University Coimbatore. He earned his Ph.D. in Computer Science from Anna University, Chennai. His prolific career includes authoring and editing over 100 books on advanced technologies and holding 27 patents. He has published over 200+ articles in esteemed journals and international conferences, including 8 papers in IEEE Transactions. As a mentor, Dr. Dhanaraj has guided four PhD candidates to completion, with eight more currently under his supervision. He is renowned for delivering insightful tech talks on disruptive technologies and has established valuable collaborations with professors from top QS-ranked universities globally. Dr. Dhanaraj's research interests include Applied AI, Cyber-Physical Systems, and Wireless Sensor Networks. His expertise in these areas has led to numerous research talks at esteemed institutions. He is a Senior Member of the IEEE, and a member of the CSTA and IAENG. Additionally, he serves as an Associate Editor and Guest Editor for several prestigious journals and is an Expert Advisory Panel Member of Texas Instruments Inc., USA.

Dr. J. Felicia Lilian, Assistant Professor, Department of Computer Science and Business Systems, Thiagarajar College of Engineering, Madurai. She has 10+ years of teaching and 2.5 yrs of research experience. She received her PhD from Anna University in the year 2023, M.E. Computer Science and Engineering from Anna University, Tirunelveli, in 2011 and a B.E. Computer Science and Engineering from Anna University, Chennai in 2009. She is a recognized supervisor from Anna University, Chennai and has scholars working on NLP and other computing technologies. She holds a membership in CSI, India. She has published 16+ articles in Scopus and SCI indexed journals, conferences and book chapter. She guided a team to the finalist in SIH 2022 and also for guided few UG students for a summer internship ICAM, Lille, France. She has been the reviewer for Journal of Supercomputing and for various reputed international conferences organized by IIIT's and NIT's. Her research interest includes Natural Language Processing, Generative AI, Deep Learning and Machine Learning.

Dr. Vandana Sharma is recognised in the Stanford list of the top 2% of scientists. She is an accomplished academician and researcher, serving as an Associate Professor at Christ University's Delhi NCR Campus, India. A Senior Member of IEEE, she has an impressive publication record, with over 100 research papers in esteemed SCI and Scopus-indexed international journals and conferences. Dr. Sharma actively contributes to the academic community through various roles, including keynote speaking, session chairing, reviewing, and participating in Technical Program Committees for renowned international journals and IEEE conferences. She holds a postdoctoral degree from Lincoln University College, Malaysia, and adjunct positions at Perdana University, Malaysia, and Beni Suef University, Egypt, where she is also a Senior Fellow of the Scientific Innovation Research Group (SIRG).

Dr. Gheorghita (George) Ghinea is a Professor in Mulsemedia Computing in the Department of Computer Science, at Brunel University London. Dr. Ghinea's research activities lie at the confluence of Computer Science, Media and Psychology. In particular, his work focuses on the area of perceptual multimedia quality and how one builds end-to-end communication systems incorporating user perceptual requirements. In the Stanford list of top 2% cited scientists in the world, Dr. Ghinea's research has been funded by both national and international organisations, consulting regularly for both public and private institutions in his areas of expertise.

Index

Also of Interest

Check out these related titles from Scrivener Publishing

Applied Computer Vision Through Artificial Intelligence, Edited by Jasminder Kaur Sandhu, Abhishek Kumar, Rakesh Sahu, and Sachin Ahuja, ISBN: 9781394272594. Master the cutting-edge field of computer vision and artificial intelligence with this accessible guide to the applications of machine learning and deep learning for real-world solutions in robotics, healthcare, and autonomous systems.

Predictive Methods in Next-Generation Computing, Edited by R. Sathiyaraj, Rajesh Kumar Dhanaraj, K. Arun Kumar, Rutvij H. Jhaveri, and A. Mohamed Abbas, ISBN: 9781394248797. Predictive Methods in Next-Generation Computing is essential for anyone looking to understand how next-generation computing technologies are driving predictive models to create smarter, safer, and more sustainable solutions across diverse fields.

Deep Learning and Its Applications Using Python, Edited by Niha Kamal Basha, Surbhi Bhatia Khan, Abhishek Kumar and Arwa Mashat, ISBN: 9781394166466. This practical book gives a detailed description of deep learning models and their implementation using Python programming relating to computer vision, natural language processing, and other applications.

Networked Sensing Systems, Edited by Rajesh Kumar Dhanaraj, Malathy Sathyamoorthy, Balasubramaniam S., and Seifedine Kadry, ISBN: 9781394310869. *Networked Sensing Systems* is essential for anyone seeking innovative and sustainable solutions across diverse sectors, as it explores the integration of cutting-edge IoT technologies and digital transformation aimed at enhancing resource efficiency and addressing climate change challenges.

Multimodal Data Fusion for Bioinformatics Artificial Intelligence, Edited by Umesh Kumar Lilhore, Abhishek Kumar, Narayan Vyas, Sarita Simaiya, and Vishal Dutt, ISBN: 9781394269938. *Multimodal Data Fusion for Bioinformatics Artificial Intelligence* is a must-have for anyone interested in the intersection of AI and bioinformatics, as it not only delves into innovative data fusion methods and their applications in omics research but also addresses the ethical implications and future developments shaping the field today.

Data Structure and Algorithms Using C++: A Practical Implementation, By Sachi Nandan Mohanty and Pabitra Kumar Tripathy, ISBN: 9781119750543. The book is intended to assist university faculty members and students to realize the principles and acquire the practical skills of programming using the C++ programming language.

Artificial Intelligence-Based System Models in Healthcare, Edited by A. Jose Anand, K. Kalaiselvi and Jyotir Moy Chatterjee, ISBN: 9781394242498. Artificial Intelligence-Based System Models in Healthcare provides a comprehensive and insightful guide to the transformative applications of AI in the healthcare system.

EXPLAINABLE MACHINE LEARNING MODELS AND ARCHITECTURES: Real-Time System Implementation, Edited by Suman Lata Tripathi and Mufti Mahmud, ISBN: 9781394185849. This cutting-edge new volume covers the hardware architecture implementation, the software implementation approach, and the efficient hardware of machine learning applications.

Natural Language Processing for Software Engineering, Edited by Rajesh Kumar Chakrawarti, Ranjana Sikarwar, Sanjaya Kumar Sarangi, Samson Arun Raj Albert Raj, Shweta Gupta, Krishnan Sakthidasan Sankaran, and Romil Rawat, ISBN: 9781394272433. Discover how Natural Language Processing for Software Engineering can transform your understanding of agile development, equipping you with essential tools and insights to enhance software quality and responsiveness in todays rapidly changing technological landscape.

MACHINE LEARNING TECHNIQUES FOR VLSI CHIP DESIGN, Edited by Abhishek Kumar, Suman Lata Tripathi, and K. Srinivasa Rao, ISBN: 9781119910398. This cutting-edge new volume covers the hardware architecture implementation, the software implementation approach, and the efficient hardware of machine learning applications with FPGA or CMOS circuits, and many other aspects and applications of machine learning techniques for VLSI chip design.

Intelligent Green Technologies for Smart Cities, Edited by Suman Lata Tripathi, Souvik Ganguli, Abhishek Kumar, and Tengiz Magradze, ISBN: 9781119816065. Presenting the concepts and fundamentals of smart cities and developing "green" technologies, this volume, written and edited by a global team of experts, also goes into the practical applications that can be utilized across multiple disciplines and industries, for both the engineer and the student.

Hybrid Intelligent Optimization Approaches for Smart Energy: A Practical Approach, Edited by Senthilkumar Mohan, A. John, Sanjeevikumar Padmanaban, and Yasir Hamid, ISBN: 9781119821243. Written and edited by a group of experts in the field, this is a comprehensive and up-to-date description of current energy optimization techniques, such as artificial intelligence techniques, machine learning, deep learning, and IoT techniques and their future trends.

NANODEVICES FOR INTEGRATED CIRCUIT DESIGN, Edited by Suman Lata Tripathi, Abhishek Kumar, K. Srinivasa Rao, and Prasantha R. Mudimela, ISBN: 9781394185788. Written and edited by a team of experts in the field, this important new volume broadly covers the design of nano-devices and their integrated applications in digital and analog integrated circuits (IC) design.

DESIGN AND DEVELOPMENT OF EFFICIENT ENERGY SYSTEMS, edited by Suman Lata Tripathi, Dushyant Kumar Singh, Sanjeevikumar Padmanaban, and P. Raja, ISBN 9781119761631. Covering the concepts and fundamentals of efficient energy systems, this volume, written and edited by a global team of experts, also goes into the practical applications that can be utilized across multiple industries, for both the engineer and the student.

Electrical and Electronic Devices, Circuits, and Materials: Technical Challenges and Solutions, edited by Suman Lata Tripathi, Parvej Ahmad Alvi, and Umashankar Subramaniam, ISBN 9781119750369. Covering every aspect of the design and improvement needed for solid-state electronic devices and circuit and their reliability issues, this new volume also includes overall system design for all kinds of analog and digital applications and developments in power systems.

MODELING AND OPTIMIZATION OF OPTICAL COMMUNICATION NETWORKS, Edited by Chandra Singh, Rathishchandra R Gatti, K.V.S.S.S.S. Sairam, and Ashish Singh, ISBN: 9781119839200. *Modeling and Optimization of Optical Communication Networks* is a comprehensive and authoritative book that delves into the optical networks, principles, technologies, and practical applications of optical networks, equipping readers with the knowledge needed to design, implement, and optimize optical networks for various applications, from telecommunications to scientific research. With its comprehensive coverage and up-to-date insights, this book serves as an essential reference in the field of optical networks.

WIRELESS COMMUNICATION SECURITY: Mobile and Network Security Protocols, Edited by Manju Khari, Manisha Bharti, and M. Niranjanamurthy, ISBN: 9781119777144. Presenting the concepts and advances of wireless communication security, this volume, written and edited by a global team of experts, also goes into the practical applications for the engineer, student, and other industry professionals.

ADVANCES IN DATA SCIENCE AND ANALYTICS, edited by M. Niranjanamurthy, Hemant Kumar Gianey, and Amir H. Gandomi, ISBN: 9781119791881. Presenting the concepts and advances of data science and analytics, this volume, written and edited by a global team of experts, also goes into the practical applications that can be utilized across multiple disciplines and industries, for both the engineer and the student, focusing on machining learning, big data, business intelligence, and analytics.

ARTIFICIAL INTELLIGENCE AND DATA MINING IN SECURITY FRAMEWORKS, Edited by Neeraj Bhargava, Ritu Bhargava, Pramod Singh Rathore, and Rashmi Agrawal, ISBN 9781119760405. Written and edited by a team of experts in the field, this outstanding new volume offers solutions to the problems of security, outlining the concepts behind allowing computers to learn from experience and understand the world in terms of a hierarchy of concepts.

MACHINE LEARNING AND DATA SCIENCE: Fundamentals and Applications, Edited by Prateek Agrawal, Charu Gupta, Anand Sharma, Vishu Madaan, and Nisheeth Joshi, ISBN: 9781119775614. Written and edited by a team of experts in the field, this collection of papers reflects the most up-to-date and comprehensive current state of machine learning and data science for industry, government, and academia.

MEDICAL IMAGING, Edited by H. S. Sanjay, and M. Niranjanamurthy, ISBN: 9781119785392. Written and edited by a team of experts in the field, this is the most comprehensive and up-to-date study of and reference for the practical applications of medical imaging for engineers, scientists, students, and medical professionals.

SECURITY ISSUES AND PRIVACY CONCERNS IN INDUSTRY 4.0 APPLICATIONS, Edited by Shibin David, R. S. Anand, V. Jeyakrishnan, and M. Niranjanamurthy, ISBN: 9781119775621. Written and edited by a team of international experts, this is the most comprehensive and up-to-date coverage of the security and privacy issues surrounding Industry 4.0 applications, a must-have for any library.

CYBER SECURITY AND DIGITAL FORENSICS: Challenges and Future Trends, Edited by Mangesh M. Ghonge, Sabyasachi Pramanik, Ramchandra Mangrulkar, and Dac-Nhuong Le, ISBN: 9781119795636. Written and edited by a team of world renowned experts in the field, this groundbreaking new volume covers key technical topics and gives readers a comprehensive understanding of the latest research findings in cyber security and digital forensics.

DEEP LEARNING APPROACHES TO CLOUD SECURITY, edited by Pramod Singh Rathore, Vishal Dutt, Rashmi Agrawal, Satya Murthy Sasubilli, and Srinivasa Rao Swarna, ISBN 9781119760528. Covering one of the most important subjects to our society today, this editorial team delves into solutions taken from evolving deep learning approaches, solutions allow computers to learn from experience and understand the world in terms of a hierarchy of concepts.

MACHINE LEARNING TECHNIQUES AND ANALYTICS FOR CLOUD SECURITY, Edited by Rajdeep Chakraborty, Anupam Ghosh and Jyotsna Kumar Mandal, ISBN: 9781119762256. This book covers new methods, surveys, case studies, and policy with almost all machine learning techniques and analytics for cloud security solutions.

SECURITY DESIGNS FOR THE CLOUD, IOT AND SOCIAL NETWORKING, Edited by Dac-Nhuong Le, Chintin Bhatt and Mani Madhukar, ISBN: 9781119592266. The book provides cutting-edge research that delivers insights into the tools, opportunities, novel strategies, techniques, and challenges for handling security issues in cloud computing, Internet of Things and social networking.

DESIGN AND ANALYSIS OF SECURITY PROTOCOLS FOR COMMUNICATION, Edited by Dinesh Goyal, S. Balamurugan, Sheng-Lung Peng and O.P. Verma, ISBN: 9781119555643. The book combines analysis and comparison of various security protocols such as HTTP, SMTP, RTP, RTCP, FTP, UDP for mobile or multimedia streaming security protocol.

SMART GRIDS FOR SMART CITIES VOLUME 2: Real-Time Applications in Smart Cities, Edited by O.V. Gnana Swathika, K. Karthikeyan, and Sanjeevikumar Padmanaban, ISBN: 9781394215874. Written and edited by a team of experts in the field, this second volume in a two-volume set focuses on an interdisciplinary perspective on the financial, environmental, and other benefits of smart grid technologies and solutions for smart cities.

SMART GRIDS AND INTERNET OF THINGS, Edited by Sanjeevikumar Padmanaban, Jens Bo Holm-Nielsen, Rajesh Kumar Dhanaraj, Malathy Sathyamoorthy, and Balamurugan Balusamy, ISBN: 9781119812449. Written and edited by a team of international professionals, this groundbreaking new volume covers the latest technologies in automation, tracking, energy distribution and consumption of Internet of Things (IoT) devices with smart grids.

Printed in the USA/Agawam, MA
September 16, 2026

910661.085